THE UNITED STATES ARMED FORCES
NUCLEAR, BIOLOGICAL AND CHEMICAL
SURVIVAL MANUAL

ALSO BY DICK COUCH

Novels

SEAL Team One

Pressure Point

Silent Descent

Rising Wind

The Mercenary Option

Nonfiction

The Warrior Elite

To Be a U.S. Navy SEAL

THE UNITED STATES ARMED FORCES NUCLEAR, BIOLOGICAL AND CHEMICAL SURVIVAL MANUAL

Everything You Need to Know to Protect Yourself and Your Family from the Growing Terrorist Threat

Compiled and Edited for Civilian Use by
Dick Couch, Captain, USNR (Ret.)

BASIC
BOOKS

A MEMBER OF THE PERSEUS BOOKS GROUP

Published by Basic Books,
A Member of the Perseus Books Group

Designed by Lovedog Studio

Text is set in Monotype Sabon

A cataloging-in-publication record for this book is available from the Library of Congress.

ISBN 0-465-00797-X

A John Boswell Associates Book

03 04 05 06 / 10 9 8 7 6 5 4 3 2 1

For the men and women serving in our military who must train, and may have to fight, in a nuclear, biological and chemical environment. Thank you for your dedication and professionalism. If the defense of our nation and freedoms takes you into harm's way, may God protect and watch over you.

Acknowledgements

I would like to thank John Boswell for his vision and Dr. Richard Rosenblatt, MD for his support and encouragement. And a big thanks to Gary Bethke and Comex Corporation for their technical assistance. There are a number who, due to security classification and job description have asked not to be acknowledged. But thanks anyway.

Contents

Introduction

You don't need to be a soldier or a politician or a police offi-
cer to understand the threat posed by nuclear, biological and
chemical weapons—not just to the United States as a nation,
but to each and every man, woman and child in our country.
You need only read the president's introductory letter in our
nation's capstone security document, *The National Security
Strategy of the United States of America,* where President
George W. Bush sums up this threat:

> The gravest danger our Nation faces is at the crossroads of
> radicalism and technology. Our enemies have openly declared
> that they are seeking weapons of mass destruction, and evi-
> dence indicates that they are doing so with determination.

This threat is so grave that, in December 2002, the White
House published, for the first time ever, *The National Strategy*

to Combat Weapons of Mass Destruction, an extraordinary document that states:

> Terrorist groups are seeking to acquire weapons of mass destruction with the stated purpose of killing large numbers of our people, and those of our friends and allies, without compunction and without warning. . . . For them, these are not weapons of last resort, but militarily useful weapons of choice.

Alternatively, one could listen to the public speeches of our political leaders. In his remarks at the U.S. Military Academy in June 2002, President Bush stated:

> Our enemies have declared the intention to attain catastrophic power to strike us. They want the capability to blackmail us, or to harm us, or to harm our friends.

Or one could listen to the speeches of our executive branch officials directly charged with the defense of the United States, such as John Bolton, Undersecretary of State for Arms Control and International Security, who declared shortly after September 11, 2001:

> There is no doubt in my mind that had the terrorists who attacked the World Trade Center and the Pentagon possessed weapons of mass destruction they would have used them in these attacks.

With this high-level attention and such strong words, it is easy and comforting to think that our government is addressing the problem—that the military will defend us or that the Department of Homeland Security will protect us. We would like to believe that the national response to the events of

September 11, 2001, has resulted in a robust defense against the terrorists who would bring weapons of mass destruction to America. I wish it were so. The United States Commission on National Security, led by former Senators Gary Hart and Warren Rudman, noted:

> A year after September 11, 2001, America remains dangerously unprepared to prevent and respond to a catastrophic terrorist attack on U.S. soil.

When I was asked to edit this book, my first reaction was "I'd like to, but I'm just too busy." I'm working on a novel that will be released by Pocket Books in late summer, with a second in the series due a year from now. The series is called *The Mercenary Option*. I'm also preparing a nonfiction work on advanced Navy SEAL training called *The Finishing School*, a sequel to *The Warrior Elite*. And I am collaborating with my photographer friend Cliff Hollenbeck, for a pictorial book on SEAL Training. This full-color, tabletop book is scheduled for release this spring. I have a lot on my plate. But I could not turn this project down. There are several reasons for this: guilt, fear, and anger.

The first is guilt. After the Berlin Wall fell and the Soviet Union left the stage as our primary adversary, novelists like me turned to terrorists for scenarios that threatened our national security. When I develop a fiction scenario, I try to think like a terrorist and I do an immense amount of research. Military and disaster-preparedness experts tell me that my plots are plausible, accurate, and scary. I've written some very realistic novels that feature chemical and radiological terror. And for well over a decade, hundreds of thousands of readers have bought these books. In these thrillers, the terrorists attempt some very nasty things and they almost succeed, but in the nick of time, the good guys come to the rescue. Now I feel a little guilty that my fiction is on the verge of becoming fact, and in real life the good guys don't always get there in time. And this gives me a great deal of concern.

Now the fear. What do you do—indeed what do I do—when the unthinkable events in my novels become fact? It can certainly happen; many experts believe it is only a matter of time. This text is a primer that can save your life and the lives of your loved ones. It is a ready resource to help you prepare for a nuclear, biological or chemical terrorist attack. The *only* antidote for fear is knowledge. Who has this knowledge? Who has been in the business for decades of preparing Americans to survive a nuclear, biological or chemical attack? There is only one answer: The United States Military.

In addition to guilt and fear, I'm also angry, and I hope you are, too. These terrorists feel that they can take advantage of our free and open society, with its guarantees of individual liberties, and use these freedoms as an avenue to attack us. Sometimes I feel like the outraged newscaster played by Peter Finch in the movie *Network:* "I'm mad as hell and I'm not going to take it anymore." But what can you and I do? I'll tell you what we can do. We can become informed, and in so doing, we can eliminate some of the fear and the potential loss of life that these terrorists wish to inflict on our wonderful nation.

We are all exposed—your family and mine. So I decided to take on this project and write something that you and your family can use. This book will quiet your fears, as it has mine, *and* it may save your life and the lives of those you care about. I have no children of my own, but many of my friends have grown kids who are off to college or working in a major city. I plan to see that each of them has one of these manuals.

To prepare this manual, I have poured through volumes of military and government agency material on nuclear, biological and chemical warfare. Consider it your "field manual" against terror. It is designed to help you and your family to survive a terrorist attack with these weapons of mass destruction. The text contains the best practices of all the United States military services for dealing with the threat of these horrifying killers. This manual is straightforward in the presentation of the facts and dangers of a terrorist attack and user-friendly in its application. I sincerely hope that all those who have bought and enjoyed my novels over the years will purchase this book. Sadly, it is time to prepare for the real thing.

One of my challenges in preparing this work was to make it a useful document with practical advice. Many of my sources were heavy on "military speak" and "medical speak," and the jargon of the professional soldiers and healers. In the book's eight chapters I have toned this down and presented the information in a way that is accessible and engaging to the public. In the appendices, however, I wanted the reader to have the information directly from the experts. I hope I have struck the right balance between the useful and the technical. The information in this book represents a best practice approach to these issues.

Keep in mind that advances in medical science and field medical procedures are taking place every day. Later editions of this manual will reflect updated information as it becomes available. Also, civilian disaster response doctrine and civil defense efforts may be more applicable than standard military procedures developed for combatant use. And finally, the clinical and medical treatments and procedures in this text are not meant as a guide for self-treatment. Just as if you are the first person at a terrible automobile accident, you have to take action to save lives until help arrives. There is no substitute for a trained medical professional. But the more you know about the proper action to take until help arrives may be critical in saving life.

I believe this field manual is your best defense against the nuclear, biological, and chemical weapons that our enemies have sworn to use against us. Keep it close by as a handy reference. I urge you to read it and use it to protect your family from the terrorist attack that we will all likely have to face in the near future.

Dick Couch
March 2003

Chapter 1

THE THREAT OF NUCLEAR, BIOLOGICAL, AND CHEMICAL TERRORISM

But even as Third Wave armies hurry to develop damage-limiting precision weapons and casualty-limiting non-lethal weapons, poorer countries are racing to build, buy, borrow, or burgle the most indiscriminate agents of mass lethality ever created; chemical and biological as well as atomic. Once more we are reminded that the rise of a new war form in no way precludes the use of earlier war forms—including the most virulent weapons.

—War and Anti-War: Survival at the Dawn of the 21st Century,
Alvin and Heidi Toffler

Though it was written in 1993, this quote by two preeminent futurists is eerily prescient of how war has dawned in this century. Nor is it an accident that the President's Advisory Panel to Assess Domestic Response Capabilities for Terrorism Involving Weapons of Mass Destruction used this quote to set the strategic context for their report.

Until the last decade of the last century, most Americans thought that terrorism was something that happened elsewhere. However frequently U.S. citizens and interests were the target of terrorists abroad, many nonetheless believed that the United States itself was somehow immune to such violence within its own borders. Terrorism, accordingly, was regarded as a sporadic—albeit attention-grabbing—problem that occasionally affected Americans traveling or living overseas and concerned only those U.S. government agencies with specific diplomatic and national security responsibilities.

If the 1993 World Trade Center bombing put a dent in that complacency and the explosion in Oklahoma City two years later added to those concerns, the attacks on the World Trade Center and the Pentagon on September 11, 2001, dramatically underscored the depth of animosity that some people harbor toward the United States. Coupled with this ill will for our country and way of life, a new terrorist mind-set seems to have emerged—one that does not shrink from indiscriminate mass murder using nuclear, biological, or chemical weapons. Indeed, today's terrorists seem determined to harm us with whatever means they have at their disposal to bring their brand of fanaticism to America.

The availability of advanced military and commercial technologies, and commonly available transportation and delivery means, permit terrorists to develop and employ weapons of mass destruction well beyond their geographic regions. Moreover, adversaries not party to an ongoing conflict may support terrorism on U.S. soil for their own purposes.

Any number of terrorist organizations or potential adversaries currently possess, or could rapidly acquire, biological and chemical weapons and other toxic materials and, in some cases, nuclear or radiological capabilities. They may also have or seek to acquire clandestine and long-range delivery systems to reach our shores. Potential enemies include emerging global adversaries, regional adversaries, and non-state adversaries. States have territorial and political bases of power, whereas non-state adversaries have population and ideological bases of power. Others rely on the shared interests and capabilities of members. Non-state adversaries include terrorist and criminal organizations and individuals with the motivation and resources to put U.S. interests at risk.

Simply put, these are people who don't understand us or like us. They are threatened by the United States because our democratic institutions and free enterprise system give rise to a middle class and place power in the hands of the people—governance with the consent of the governed. They envy our wealth and freedoms, and they fear our cultural influence. Since they are no match for our military power, they resort to terror; and since they have no hope of defeating our soldiers, they attack our civilians—they attack you.

States may have incentives to acquire weapons of mass destruction in spite of their adherence to international agreements and treaties forbidding such actions. Non-state groups do not consider themselves bound by such agreements and treaties. States and non-state groups alike may have incentives to operate outside the norms of acceptable international behavior, especially when important interests are involved. They may seek to overcome U.S. alliance or coalition strengths by focusing on vulnerabilities to weapons of mass destruction. There are at least five principal reasons why terrorists resort to using weapons of mass destruction as a weapon.

The first reason, and the most basic, may be simply the desire to kill as many people as possible. Weapons of mass destruction could give a terrorist group the potential capability to wipe out thousands, possibly even hundreds of thousands, in a single stroke. Statistics provided by the Federal Emergency

Management Agency (FEMA) give one indication of the potential killing power of these agents compared to conventional high explosives. According to FEMA, to produce about the same number of deaths within a square mile, it would take 705,000 pounds of fragmentation cluster bomb material; 7,000 pounds of mustard gas; 1,700 pounds of nerve gas; 11 pounds of material in a crude nuclear fission weapon; 3 ounces of botulinal toxin type A; or half an ounce of anthrax spores. Such weapons would provide terrorists with the perfect means to seek revenge against, or even annihilate, their enemies.

Another reason for groups to seek the use of weapons of mass destruction is to exploit the classic weapon of the terrorist—fear. Terrorism is a form of psychological warfare. The ultimate objective is to destroy the structural supports that give society its strength by showing that the government is unable to fulfill its primary security function and, thereby, eliminating the solidarity, cooperation, and interdependence on which social cohesion and economic functioning depend. Viewed in this context, even a limited attack using weapons of mass destruction would have disproportionately large psychological consequences, generating unprecedented fear and alarm throughout society. The 1995 nerve gas attack by the Aum Shinrikyo cult, for instance, which resulted in 12 deaths, not only ignited mass panic in Tokyo, but also shattered the popular perception of security among the Japanese people, who had considered their country to be among the safest in the world. Moreover, it served to galvanize American attention to the threat of weapons of mass destruction.

A third rationale for why terrorists resort to weapons of mass destruction could be the desire to negotiate from a position of unsurpassed strength. A credible threat to use nuclear, biological, or chemical weapons would be unlikely to go unanswered by a government and could, therefore, provide a terrorist organization with a powerful tool of political blackmail of the highest order.

An additional reason why terrorists would use weapons of mass destruction—and especially biological weapons—could derive from certain logistical and psychological advantages that such weapons might offer terrorists. A biological attack, unlike a conventional bombing, may or may not be likely to attract immediate attention, and could initially go unnoticed, only manifesting itself days or even weeks after the event. This would be well suited to groups that wish to remain anonymous, either to minimize the prospect of detection and arrest or to foment greater insecurity in their target audience by appearing as enigmatic, unseen, and unknown assailants.

Finally, a terrorist group may wish to use weapons of mass destruction—and especially biological agents—to cause economic and social damage by targeting a state's or region's agricultural sector. On several previous occasions in other parts of the world, terrorists have contaminated agricultural produce or have threatened to do so. Between 1977 and 1979, more than 40 percent of the Israeli European citrus market was jeopardized by a Palestinian plot to inject Jaffa oranges with mercury. In 1989, a Chilean radical left-wing group that was part of an anti-Pinochet movement claimed that it had laced grapes bound for U.S. markets with sodium cyanide, causing suspensions of Chilean fruit imports by the United States, Canada, Denmark, Germany, and Hong Kong. In the early 1980s, Tamil separatists in Sri Lanka threatened to infect Sri Lankan rubber and tea plantations with nonindigenous diseases as part of a biological war strategy designed to cripple the Sinhalese-dominated government. The United States is vulnerable to this form of aggression, given the integrated and intensive nature by which farm animals are bred, transported, and sold, as well as the high degree of genetic homogeneity and concentration found in primary crop-growing regions. Disrupting this vital and vulnerable industry could not only damage the economy, it could also undermine confidence in the government's ability to protect the very foundations of American society.

Preparedness at home plays a critical role in combating terrorism by reducing its appeal as an effective means of warfare. Acts of catastrophic terrorism produce not only deaths and physical destruction but also societal and economic disruption. So, it is important to try to eliminate the incentive for undertaking acts of terrorism on U.S. soil in the first place. If the disruptive effects of terrorism can be sharply reduced, America's adversaries may be deterred from taking their battles to the streets of our nation's homeland.

The threat of terrorists using nuclear, biological, and chemical materials appears to be rising—particularly since the September 11, 2001, attacks. Several of the 30-plus identified foreign terrorist organizations and other non-state actors worldwide have expressed interest in these weapons—although terrorists probably will continue to pursue proven conventional tactics such as bombings and shootings.

Nuclear, biological, and chemical weapons information and technology are widely available in text form or on the Internet. Also, the extensive publicity surrounding the anthrax incidents following the September 11, 2001, attacks has highlighted the vulnerability of civilian and government targets to chemical, biological, radiological, and nuclear attacks.

Although al-Qaeda and other terrorists will continue to use conventional weapons, one of the highest concerns is their stated readiness to attempt unconventional attacks against the United States. As early as 1998, Osama bin Laden publicly declared that acquiring unconventional weapons was "a religious duty."

Terrorist groups around the world have ready access to information on chemical and biological and, to some extent, even nuclear weapons via the Internet, publicly available scientific literature, and scientific conferences, and we know that al-Qaeda was working to acquire some of the most dangerous chemical agents and toxins. A senior bin Laden associate on trial in Egypt in 1999 claimed that his group had chemical and biological weapons. Documents and equipment recovered from al-Qaeda facilities in Afghanistan show that bin Laden has a more sophisticated biological weapons research program than previously known.

There is strong evidence that al-Qaeda has ambitions to acquire or develop nuclear and radiological weapons and has been receptive to any outside nuclear assistance. In February 2001, during the trial on the al-Qaeda bombings of the American embassies in Tanzania and Kenya, a government witness—Jamal Ahmad Fadl—testified that al-Qaeda pursued the sale of a quantity of purported enriched uranium (which in fact probably was scam material) in Sudan in the early 1990s. On the basis of the materials that known terrorist groups have attempted to acquire, the Central Intelligence Agency assesses the terrorist use of radiological dispersal devices to be a highly credible threat.

This near-term threat of nuclear/radiological, biological, or chemical terrorism was underscored by Secretary of Defense Donald Rumsfeld, who, shortly after the events of September 11, 2001, noted:

In World War II there were suicide pilots flying their aircraft into our ships. Today, a new enemy is seeking global power and has flown our airliners into our buildings on suicide missions. They are now working to gain access to weapons of mass destruction. . . and armed with these weapons they will kill not a few hundred or a few thousand, but tens of thousands or hundreds of thousands.

It would be a grave mistake to think that the threat of terrorists using weapons of mass destruction is a hypothetical. In early 2003 six men were arrested in London, England, with a quantity of the deadly toxin ricin in their possession. This is only one of an alarming series of similar incidents, pointing out the diffuse nature of the threat of these weapons. Unfortunately, many prefer to think of this threat as more hypothetical than real. George Poste, chairman of the Department of Defense bio-terrorism task force, addressed this in a very blunt and direct manner:

> I do not wish to see the coffins of my family, my children, and my grandchildren created as a consequence of the utter naïveté, arrogance and hubris of people who cannot see there is a problem.

The threat of nuclear, biological, and chemical terrorism is real and it is growing—and this threat requires that all of us take proactive steps to protect our families and ourselves. You can start with a comprehensive family action plan.

Chapter 2

A COMPREHENSIVE FAMILY ACTION PLAN

2.1   Preparing for Nuclear, Biological, and Chemical Terrorist Attacks

2.2   Developing a Family Emergency Action Plan

2.3   Checklists for Disasters Including Terrorist Attacks

2.4   Special Procedures for Nuclear, Biological, and Chemical Attacks

The need to prepare for disasters did not begin on September 11, 2001, nor will it end when the current terrorist threat is dealt with. Citizens and governments have had to deal with disasters for all of recorded history. Exactly forty years ago, the Office of Civil Defense within the U.S. Department of Defense published the *Personal and Family Survival Manual* as the authoritative text for a Civil Defense Adult Education Course. In this compact manual, the authors quote then-Secretary of Defense Robert McNamara, who noted that: "In some wartime situations a reasonable Civil Defense program could do more to save lives than many active defense measures."

Secretary McNamara's quote was focused on the then compelling threat of nuclear weapons and took place in an era when every major city had conspicuous signs leading to community fallout shelters. Tens of thousands of citizens constructed their own family shelters. I remember grade-school drills where we crawled under our desks to prepare for a nuclear blast. The need for citizens today to take an active role in preparing for terrorist attacks with weapons of mass destruction is no less compelling than the threat of nuclear annihilation a generation ago. In fact, I would submit that the threat of attack is higher today than during the Cold War with the prospect of Soviet atomic attack. **A Family Emergency Action Plan will do more to save lives than almost any activity carried out by the government.**

2.1   Preparing for Nuclear, Biological, and Chemical Terrorist Attacks

Americans are now more conscious of terrorism and the potential for terrorist attacks than they have been at any other time in our history. **The threat is real, and the threat is growing.**

If you are like the majority of your fellow citizens you are probably concerned about this threat but are not at all certain what you should do about it. You are not alone. Some people think about the two extreme ends of the spectrum: (1) the morbid fear that a terrorist attack with weapons of mass destruction might happen at any time, and (2) the "devil-may-care" attitude that if you don't worry about it, nothing bad will happen, or, it could happen—but not to me. Neither end of the spectrum represents a rational attitude. As for an attack happening to me or in my community, ask any New Yorker about that.

Living in morbid fear of terrorism and putting your life on hold while you worry about it would constitute a victory for these terrorists. They have caused you to change your way of life—*our* way of life—and sign away part of your freedoms, liberties that this nation has stood for since its founding over two centuries ago. But to operate at the other end of the spectrum—not giving any thought to what could happen if there was a terrorist attack with weapons of mass destruction—is, plainly stated, foolhardy.

Most people are in the middle of this spectrum: They want to do something actively to prepare but they are not exactly sure what to do. Reading this "field manual" is an important first step, because what I said at the outset of this book is true—the antidote for fear is knowledge. By reading this book you have taken a significant first step in acquiring the knowledge you need to help protect you and your family from an all-too-probable terrorist attack with weapons of mass destruction.

There are commonsense actions that *all* citizens can take to assure that in the event of a terrorist attack they are as prepared as possible to survive. However, the spectrum of threats we could face goes beyond just terrorist threats. The threat from floods, fires, earthquakes, hurricanes, tornadoes, and other natural and man-made disasters is always with us.

Helping you deal with each and every one of these potential disasters is beyond the scope of this book. We want to focus on the very real threat of a terrorist attack with weapons of mass destruction. The initial steps in the commonsense preparations a person would make to cope with other types of disasters closely parallel those that you would take to prepare for a terrorist attack; this is dual-use information.

These steps were developed by the government experts who are charged with leading the nation in dealing with disasters—the Federal Emergency Management Agency, or FEMA. Soon, this information will be augmented by directives from the Department of Homeland Security, but it may be a while before it is available to the public. FEMA has developed these tactics, techniques, and procedures based on their experience in dealing with literally thousands of disasters across the entire array of natural and man-made catastrophes. Thus, we present our Comprehensive Family Action Plan within the context of what FEMA has taught us to do in other disaster scenarios. Our goal is for you to treat the threat of a terrorist attack with weapons of mass destruction as one part of your action plan. Take charge and make it happen! You will feel much better for having done so.

2.2　Developing a Family Emergency Action Plan

Checklists are vital. Military aviators and airline pilots know that using such checklists are absolutely crucial to their safety and survival. It's human nature to resist anything that resembles following orders or established procedure. We have busy lives and jobs; we have many things we *have* to do. Taking the time to do some emergency planning can easily be overtaken by recreation or family time or just watching TV. And some of us are just plain lazy—"I'll do it tomorrow."

However, when disaster strikes there is *absolutely no time* to prepare or to find out what to do. You must have a plan, you must have rehearsed that plan, and you must have acquired any equipment and material you will need in a disaster *before* it happens. There is a name for people who fail to do these things. They are called "victims" or "fatalities." Don't be one of them.

These checklists are guidelines—and every person will need to adapt these checklists to their own individual circumstances and those of their family. The point is that by failing to plan you are essentially planning to fail. The specter of a terrorist attack with weapons of mass destruction may seem just too real and too terrifying to even have a plan. Not so! The checklists presented in the following pages begin with commonsense preparations for all types of emergency situations and then move to some specific guidance for nuclear, radiological, biological, and chemical terrorist attacks.

2.3   Checklists for Disasters Including Terrorist Attacks

These checklists were developed primarily from those provided by FEMA as well as from military experience with tactics, techniques, and procedures used during emergencies. Beyond checklists, however, you must keep in mind the "big picture" of what you and your family must do in the event of any disaster—but especially in dealing with the chaos that surely will follow any terrorist attack. These checklists are a guide; use common sense to adapt them to your own individual and family situations. The three critical pieces to any Family Emergency Action Plan are:

Communication: How will your family members communicate with one another?

Supplies: What supplies must you have on hand if you remain in your home or flee?

Destination: Are you going to your safe-room or to another destination?

Family Emergency Action Plan

Your family must have an emergency plan to deal with all types of disasters. Most of these recommended actions seem intuitive. You are likely to read them and tell yourself, "Sure, we'll do that." Unfortunately, experience with *victims* of disasters shows that all too often good intentions are not followed through with effective action. Don't be one of those victims. Develop your own plan for *your family*. Start with these useful guidelines:

❋ Meet with all family members to discuss the dangers of all kinds of emergencies.

❋ Together, develop a family action plan to deal with each particular emergency.

❋ Find the safe spots in your home for each type of disaster.

❋ Discuss what to do about power outages and personal injuries.

❋ Draw a floor plan of your home and mark at least two escape routes from each room.

❋ Show family members how to turn off the water, gas, and electricity at the main switches.

❋ Post emergency telephone numbers at telephones and have printed copies with cell phones.

* Teach children how and when to call 911, police, and fire.

* Instruct household members to turn on the radio for emergency information.

* Pick one local and one out-of-state friend to call if separated during a disaster.

* Pick two emergency meeting places.

  + One place near your home in case of a fire.

  + One place outside your neighborhood in case you cannot return home.

* Take a basic first aid and CPR class; buy a Boy Scout manual.

* Keep family records in a waterproof and fireproof container.

Finally, should the disaster require you to stay inside of your home to avoid external hazards, go to your safe-room and bring all needed supplies with you. This safe-room should be a room without windows if possible, and should contain a telephone and a radio. Be aware that in the event of a chemical attack, heavy chemical vapors sink to the lowest point in the house. For this reason, basements would not be optimal for safe-rooms. However, as will be explained in Chapter 4, in the event of a nuclear or radiological attack, a basement is the most preferred area of the house.

Family Emergency Evacuation and Escape Plan

There will be some emergencies where you may need to evacuate your house or apartment at a moment's notice. You must be ready to get out fast. Develop an escape plan by drawing a floor plan of your residence. Using a black or blue pen, show the location of doors, windows, stairways, and large furniture. Indicate the location of emergency supplies (Family Disaster Supplies Kit, Family Car Getaway Kit), fire extinguishers, smoke detectors, collapsible ladders, first aid kits, and utility shutoff points. Next use a colored pen to draw a broken line charting at least two escape routes from each room. Finally, mark a place outside of the home or apartment house where household members should meet in case of evacuation. Be sure to include important points outside such as garages, patios, stairways, elevators, driveways, and porches. If your home has more than two floors, use an additional sheet of paper. Practice emergency evacuation drills with all household members at least two times each year.

Armed with this expanded floor plan, when the time comes to evacuate your house, take the following steps:

* Listen to a battery-powered radio for the location of emergency shelters.

* Follow the instructions of local officials.

* Wear protective clothing and sturdy shoes.

* Take your Family Disaster Supplies Kit.

* Lock your home.

* Use travel routes specified by local officials.

✳ Have cash on hand; the ATMs may not be operable.

Once these steps are completed, and only if you have time without putting your family in jeopardy:

✳ Shut off water, gas, and electricity, if instructed to do so.

✳ Let others know when you leave and where you are going.

✳ Make arrangements for pets—animals are not allowed in shelters.

Family Home Hazard Checklist

Many seemingly benign items that we have in our homes on a day-to-day basis can become deadly items in a disaster, particularly if your home is subjected to extreme shocks that would occur in a nuclear or radiological attack. These ordinary items can cause injury and damage. Anything that can move, fall, break, or cause a fire is a potential hazard. Take the following steps *well before* a disaster:

✳ Repair defective electrical wiring and leaky gas connections.

✳ Fasten all shelves securely to a sturdy wall.

✳ Place large, heavy objects on lower shelves.

✳ Hang pictures and mirrors away from beds.

✳ Brace overhead light fixtures.

✳ Secure the water heater by strapping it to wall studs.

* Repair cracks in the ceiling and foundation.

* Store weed killers, pesticides, and other hazardous items away from heat sources.

* Place oily polishing rags or waste in covered metal cans.

* Clean and repair chimneys, flue pipes, vent connectors, and gas vents.

Family Disaster Supplies Kit

The contents of your Family Disaster Supplies Kit will vary significantly based on the size of your family and the special needs of individual family members. However, there are a number of standard items that should form the basis of all Family Disaster Supplies Kits. Disaster preparedness experts indicate that many families preparing for an emergency typically underestimate the amount of water they need and often fail to make adequate provisions to have enough needed prescription medicine on hand. Start your Family Disaster Supplies Kit with the following items:

* At least one gallon of water per person per day—store in unbreakable containers

* Prescription medication to last at least a week—more if traveling a long distance

* A supply of nonperishable packaged or canned food and a hand-operated can opener

* Special foods as needed for infants, elderly persons, or persons on special diets

* Nonprescription drugs that might be needed, such as aspirin, antacids, laxatives, etc.

* A change of clothing, rain gear, and sturdy shoes for each family member

* Thermal blankets and sleeping bags

* A first aid kit that includes extra items over and above a normal workplace first aid kit

* Sanitation and personal hygiene items

* A cellular telephone along with a list of important phone numbers

* A battery-powered radio, flashlight, and plenty of extra batteries and bulbs

* Credit cards and cash

* An extra set of car keys

* A list of family physicians with phone numbers

* Important family documents, such as passports, wills, and financial records

* "Outdoors" items, such as compass, matches, tent, plastic containers, etc.

* Items you feel you may want to include for *your* kit; a Gameboy for the kids, a Bible or other reading

Family Car Getaway Kit

In addition to the items in the portable Family Disaster Supplies Kit, you should assemble an emergency Family Car

Getaway Kit in the event you and your family need to evacuate your home and move to another location. This kit should include:

* Powerful flashlight and/or safety lantern with plenty of batteries

* Booster cables

* Fire extinguisher

* Tire repair kit and pump or aerosol flat-tire device

* Maps, flares, and other emergency signaling devices

* Full gas tank

* Certified container for gasoline

2.4   Special Procedures for Nuclear, Biological, and Chemical Attacks

The procedures and checklists in the preceding section are general or "omnibus" procedures that will help your family get through any significant emergency. There are additional procedures unique to nuclear, biological, and chemical attacks—and especially the latter two. These suggested procedures are presented in the following paragraghs.

Much media attention has been focused on inoculations that can be given as a defense against biological or chemical agents dispersed in a terrorist attack. Anthrax and smallpox are the two pathogens that have garnered the most attention. Additionally, there has been a great deal of public anxiety regarding vaccines and antibiotics that can be administered after a biological or chemical attack.

Few vaccines come without the risk of significant side effects, and therefore public health officials and responsible physicians do not recommend wholesale public inoculations as a means of protecting the public from these pathogens *unless there is significant evidence that an attack may be imminent.* Therefore we do *not* recommend that citizens be inoculated against these pathogens absent evidence of a compelling need. Nor do we recommend that people stockpile antibiotics for use after a biological or chemical attack. Antibiotics have a limited shelf life and it is probable that they would expire before you need them. In short, dealing with biological and chemical pathogens is a job for professional first responders and medical personnel.

What citizens *can* do, by reviewing Appendices B and C of this manual, is to learn what pre- and post-attack vaccines are available for each biological or chemical agent. Then, armed with this information, they should contact their personal physician to inquire as to where they would go to receive these inoculations should an attack with biological or chemical agents take place. Most communities have disaster preparedness procedures with designated locations set aside for people to receive vaccinations or antibiotics against biological agents. This is the knowledge you will need in the event of an imminent or actual attack.

In the event of an impending or actual terrorist attack with weapons of mass destruction, consider the following:

* **Take this manual with you!** It will serve as a valuable reference.

* Follow the directions of emergency response personnel, the military, and National Guard.

Special Procedures
When Away From Home

If you are away from home, and warning of an attack is given, proceed home quickly unless otherwise directed by the authorities to remain where you are.

Also, if you are away from home during the warning of an attack, avoid high-profile venues.

If you are outside during an actual attack, cover all exposed skin surfaces, and protect the respiratory system as much as possible by covering your mouth and nose and shield your eyes. Wetted towels or paper towels may be your best available protection.

If the chemical or biological incident occurs inside a building, leave the building immediately and try to avoid the contaminated area on your way out.

Special Procedures When at Home

If there is a terrorist event requiring you to seek shelter in your home, gather all family members and all emergency supplies and proceed to your safe room.

Close all doors and windows and turn off the ventilation system. If you have not already done so, seal all windows and doors with plastic tape.

Attempt to learn what kind of attack has occurred in order to anticipate needed first aid.

Procedures When Your
Children Are at School

No one wants to scare their children or create an atmosphere of fear, but think how worried *you* will be should there be a nuclear, chemical, or biological attack and your kids are at school. Talk to them. Make a plan. Think about what you will do. Talk to your school authorities and find out what they will do in the event of such an attack. Jumping in the car and racing off to find your children could be the worst course of action for you and your kids. Along these same lines, plan what will happen if you are home and your spouse is at work.

As with all checklists, make sure these are adapted to your own family situation and your own community. And keep them updated. As we move through this era of terrorism, new information and procedures will be developed as time goes on. Preparation for a terrorist attack is a growth industry. Be sensitive and responsive to new information and keep your checklists current. Seek out other available information like the CERT (Community Emergency Response Team) and American Red Cross programs. The Internet is a wonderful source of information, but only if you use it *before the terrorist attack and write that information down in an easy-to-access place.*

Dear Reader:

I fully understand that these checklists are rather intimidating. I also understand that you may be reluctant to take these precautions or even run family disaster-preparedness drills. You may even feel a little foolish. And sadly, no one else in your neighborhood may be doing this. It's hard; this is the MTV generation. Life is good in America, even the post–9/11 America. Why scare the kids, right? Consider this: Most parents have cautioned their children about drugs, getting into a car with a stranger, or unprotected sex. Losing a child to one of these societal problems would be unbearable, but think of your guilt if you had not counseled your child about these dangers—that it could have been prevented if you had taken action before it happened. Be honest, you would feel like a criminal— for the rest of your life. Terrorism on a mass scale is now one of *our* societal problems. It's up to you; how much guilt can you handle.

Focusing only on today's headlines could lead one to think that the use of weapons of mass destruction is a new phenomenon. Nothing could be further from the truth as we'll see in Chapter 3. Although American civilians are more at risk than ever before, these weapons of terror have been used—oftentimes with stunning success—for more than two millennia.

Chapter 3

A BRIEF HISTORY OF NUCLEAR, BIOLOGICAL, AND CHEMICAL WARFARE

Gas! GAS! Quick boys. . . an ecstasy of fumbling. . . fitting the clumsy helmets just in time. . . but someone still was yelling out and stumbling. . . dim, through the misty panes and thick green light. . . as under a green sea I saw him drowning. In all my dreams, before my helpless sight, he plunges at me, guttering, choking, drowning.

—Lieutenant Wilfred Owen, British Royal Army,
excerpted from "Dulce et Decorum Est"

Disease has been part of war throughout history. The hardships of fatigue, famine, and malnutrition to which soldiers as well as civilians are subjected by war make them extraordinarily susceptible to infectious diseases. Historical records indicate that humans attempted to employ biological agents in warfare long before there was any factual knowledge of the causes of disease. The earliest attempts were probably the ancient practices of poisoning water supply points with the bodies of disease victims, leaving diseased bodies in areas the enemy was expected to occupy, or catapulting the bodies of disease victims over the walls of besieged cities. However, despite historical allusions to these practices, there are few references to their employment in specific battles or campaigns in early times, probably because such practices were not considered to be heroic and so were not glorified in songs and stories of manly combat.

The two earliest recorded uses of biological weapons in warfare occurred in the sixth century B.C., with the Assyrians poisoning enemy wells with rye ergot, and the Athenian lawgiver and poet Solon using the purgative herb hellebore during the siege of Krissa. While other instances of this type of warfare must have occurred, romanticized battle songs and stories left them unrecorded.

The use of chemical weapons dates from at least 423 B.C. when allies of Sparta in the Peloponnesian War took an Athenian-held fort by directing smoke from lighted coals, sulfur, and pitch through a hollowed-out beam into the fort. Other conflicts during the succeeding centuries saw the use of smoke and flame, and the Greeks invented Greek firs, a combination of rosin, sulfur, pitch, naphtha, lime, and saltpeter. This floated on water and was particularly effective in naval operations.

By the Middle Ages, recorded history moved beyond songs and stories and with this came factual accounts of the use of disease in warfare. It is likely that the epidemic of bubonic plague, the "Black Death" that ravaged Europe from 1347 to 1352, was the outgrowth of an early crude attempt at biological warfare. In the fourteenth century European traders were continually harassed by Tartar warriors from central Asia. Many of them fled with their merchandise to Kaffa (the present-day city of Feodosia), a small, fortified post opposite the Crimean Peninsula. Here they were besieged for nearly three years by the

Tartars. During that time, conditions in Kaffa deteriorated and were only occasionally relieved by the arrival of supply ships. Then the plague broke out among the besieging Tartars and many of them died. The Tartars catapulted these corpses over the walls of the fort, and the plague soon spread within. The plague took such a toll on the Tartars, however, that they lifted the siege of Kaffa in 1347 and dispersed. Many of the traders escaped Kaffa in a ship bound for Genoa. Soon after the ship reached the city, plague broke out in Genoa and rapidly spread throughout the continent. Four centuries later, Russian troops used the same tactic against Sweden in 1710, casting plague-infected bodies into Swedish-held Reval, Estonia.

Another early attempt to employ disease as a weapon occurred in the fifteenth century when Francisco Pizarro is said to have presented South American natives with variola-contaminated clothing that spread smallpox rapidly within the Indian population. A second, well-documented case occurred during the French and Indian War. Angered by the transfer of their lands to the king of England after the English defeat of the French in North America in 1759, Pontiac, chief of the Ottawa Indians, conceived a plot to expel the British. Pontiac enlisted the support of almost every Indian tribe from Lake Superior to the lower Mississippi for this effort. Of twelve fortified posts attacked, eight were captured and the garrisons massacred. Several relief expeditions were nearly annihilated, and many settlements were plundered and destroyed.

These successful large-scale Indian attacks drove the British to resort to desperate measures. The commander-in-chief of the British forces was aware that smallpox, which was unknown in America before the arrival of Europeans, had frequently decimated Indian tribes. He directed his subordinate, the commander of the First Battalion of Royal Americans defending Fort Pitt, to give the Indians besieging the fort blankets as part of a peace offering—blankets that had been taken from the fort's smallpox hospital. Within a few months, smallpox was prevalent among the Indian tribes of the Ohio River valley and their direct threat to British forces dissipated.

A century later during the U.S. Civil War, biological agents were employed by the Confederate forces during the Union campaign against Vicksburg in the summer of 1863. When Confederate reinforcements intended for Vicksburg were turned back and pursued by the Union forces, the retreating armies delayed their pursuers by driving sick farm animals into ponds and shooting them, thus polluting the water supply.

The beginning of the twentieth century saw a dramatic rise in the use of biological and chemical agents in warfare. The first modern attempts at biological warfare were made by Germany during World War I. The disease known as glanders was widespread in Europe during this war, and more than 50,000 horses in the French Army alone were infected with it. Both the acute and chronic forms of the disease have relatively high mortality rates when untreated, but even in nonfatal cases the animals are rendered unfit for service. During World War I, the Germans attempted to spread glanders by inoculating Romanian cavalry horses with the bacteria that causes the disease. German agents in the United States even inoculated animals scheduled for shipment to Europe with various bacteria in the hope that they would spread disease upon their arrival.

However, it was chemical warfare that most devastatingly characterized World War I. On April 22, 1915, near Ypres, Belgium, German troops dispensed an estimated 150 tons of chlorine gas from some 6,000 cylinders and this gas drifted into Allied trenches. The gas caused over 5,000 casualties to the approximately 15,000 Allied troops in the area. Had the Germans used more chlorine, the results could have been even more pronounced. After these attacks, the Allies developed protective respirator masks that provided complete protection against chlorine. Six months after the German attack at Ypres, the British retaliated with a chlorine attack at Loos.

Chlorine was not the only toxic chemical employed during World War I. During 1916 and 1917, the principal toxic chemical agents that were used included lung irritants or choking agents, such as phosgene and chloropicrin. All had to be inhaled to have an effect. By the summer of 1917, protective masks with such colorful names as Black Veil Respirator, Flannel Hood, Tissot Mask, Horse Mask, and British Box Respirator had been so improved that they furnished full protection against those agents.

To break the deadlock, the German Army introduced a new type of toxic chemical agent—mustard, a highly effective vesicant or blister agent. On July 12, 1917, again near Ypres, German artillery shells delivered this new mustard agent. That attack alone resulted in 20,000 Allied casualties. The development of mustard meant that a protective mask no longer provided total protection. To be completely safe from mustard, the entire body had to be protected. The enemy's introduction of mustard came as a complete surprise to the Allies and caused tens of thousands of casualties before any form of defense could be devised. In the summer of 1917 alone over 14,000 mustard-poisoning cases were admitted to British casualty clearing stations. During the remainder of the war, mustard was used extensively by both sides. It was the greatest single casualty producer of all weapons used during the period. These vesicants caused 400,000 casualties, nearly one-third of the entire war's total chemical casualties, even though the vesicants used in battle accounted for only one-tenth of all chemical agents used throughout the war. During the war, most of the combatants developed protective dress and masks to deal with these chemicals. The Russians alone failed to take defensive measures and suffered more casualties that all other combatants combined.

The military use of chemical agents continued after the end of World War I. Great Britain used chemicals against Russians and mustard against the Afghans north of the Khyber Pass, and Spain employed mustard shells against the Riff tribes of Morocco. During the next decade the Soviet Union used lung irritants against tribesmen in Kurdistan and Italy used tear gas and mustard during the war against Abyssinia (now Ethiopia).

The years leading up to World War II saw extensive research into chemical and biological weapons by almost all of the eventual Allied and Axis combatants. It is intriguing that these weapons were not used extensively during this war. Historians suspect that the advent of air power—which enabled the combatants to rain chemical or biological destruction down on each other from the air—may have mitigated their use. Indeed, this may have been the first instance of mutually assured destruction that prevented the use of an available weapon of mass destruction.

Nonetheless, all the combatants maintained active chemical and biological research programs throughout the war and substantial quantities of agents were produced. As they had in World War I, the Germans led the way in this research and production, producing a much more potent toxic chemical agent, tabun, the first of the nerve agents. Unlike earlier toxic chemicals, a comparatively small dose of nerve agent is required to produce casualties. Over a half century after Germany introduced this deadly chemical, nerve agents like tabun remain the most deadly toxic chemical agents known today.

Japan did use biological agents in World War II to a limited extent. The earliest recorded instances of the Japanese attempts to produce biological agents occurred in 1937 when that nation started an ambitious biological warfare program 40 miles south of Harbin, Manchuria. This infamous laboratory complex was known by its code name, "Unit 731." Studies directed by Japanese General Ishii Hiro continued until 1945, when the complex was burned. A post–World War II investigation revealed that the Japanese conducted research on numerous organisms and used Chinese prisoners of war as research subjects. Nearly 1,000 human autopsies apparently were carried out at Unit 731, mostly on victims exposed to aerosolized anthrax. Many more prisoners and Chinese nationals may have died in this facility—some historians have estimated up to 3,000 deaths. Following reported overflights by Japanese planes suspected of dropping plague-infected fleas, a plague epidemic broke out in China and Manchuria. Survivors from Changde, Zhejiang Province, have testified that flea-infested grain dropped from Japanese planes claimed the lives of 6,491 Chinese in that city alone. By 1945, the Japanese program had stockpiled 800 pounds of anthrax to be used in specially designed fragmentation bombs.

The United States became the first—and to date the only— nation to use nuclear weapons in wartime when it dropped two nuclear weapons on Japan in August 1945, one on the city of Hiroshima and, three days later, one on the city of Nagasaki. Taken together, these bombs killed approximately 100,000 people and led directly to the unconditional surrender of the Japanese nation. One of these nuclear bombs produced the same effect of 12,700 tons of TNT and the ground temperature immediately beneath the explosion was estimated to be at least 5,000 degrees Centigrade.

Biological weapons have been a cause of particular concern since World War II, with several cases of suspected or actual use. Among the most widely reported were the use of ricin as an assassination weapon in London in 1978, and the accidental release of anthrax spores at Sverdlovsk in 1979.

In 1978, a Bulgarian exile named Georgi Markov was attacked in London with a device disguised as an umbrella. The device injected a tiny pellet filled with ricin toxin into the subcutaneous tissue of his leg while he was waiting for a bus. He died several days later. During a subsequent autopsy, the tiny pellet was found and determined to contain the toxin. It was later revealed that the Bulgarian Secret Service carried out the assassination, and the technology to commit the crime was supplied by the former Soviet Union.

In April 1979, an incident occurred in Sverdlovsk (now Yekaterinburg) in the former Soviet Union, which appeared to be an accidental aerosol release of *Bacillus anthracis* spores from a Soviet military microbiology facility, Compound 19. Residents living downwind from this compound developed high fever and difficulty breathing, and a large number died.

The use of chemical agents as a weapon of war in the last half of the twentieth century was less prevalent than the use of biological weapons. It has also been reliably reported that in 1983 Iraq used mustard agents as well as the nerve agent tabun against Iranian troops. Later in the war, Iraq apparently began to use the more volatile nerve agent sarin, and Iran may have used chemical agents in an attempt to retaliate for Iraqi attacks.

In December 1990, the Iraqis filled 100 R400 bombs with botulinum toxin, 50 with anthrax, and 16 with aflatoxin. In addition, 13 Al Hussein (SCUD) warheads were filled with botulinum toxin, 10 with anthrax, and two with aflatoxin. These weapons were deployed in January 1991 to four locations. In all, Iraq produced 19,000 liters of concentrated botulinum toxin (nearly 10,000 liters filled into munitions), 8,500 liters of concentrated anthrax (6,500 liters filled into munitions), and 2,200 liters of aflatoxin (1,580 liters filled into munitions).

In August 1991, the United Nations carried out its first inspection of Iraq's biological warfare capabilities in the aftermath of the Gulf War. In the course of this inspection, representatives of the Iraqi government announced to leaders of the United Nations Special Commission that they had conducted research into the offensive use of *Bacillus anthracis, Clostridium botulinum*, and *Clostridium perfringens*. This admission of biological weapons research verified many of the concerns of the U.S. intelligence community. Iraq had extensive and redundant research facilities at Salman Pak and other sites, many of which were destroyed during the war.

In 1995, further information on Iraq's offensive program was made available to United Nations inspectors. Iraq admitted to conducting research and development work on aflatoxins, wheat cover smut, and ricin. Field trials were conducted with *Bacillus subtilis* (a stimulant for anthrax), botulinum toxin, and aflatoxin. These agents were also tested in various delivery systems, including rockets, aerial bombs, and spray tanks.

Many lessons can be learned from past warfare involving weapons of mass destruction and the U.S. experience in combating these weapons. The relatively few anthrax-tainted letters mailed in the United States in late 2001 caused the evacuation of government offices and crippled our postal system for a period of time. Thus far, the United States has been extremely lucky and has not experienced a chemical, biological, or radiological "Pearl Harbor." To guard against this eventuality, the United States must continue to investigate this type of warfare and its forces—military, civilian, first responders, and others— must be prepared to deal with these weapons should they be released. This includes you, your family, and your friends and neighbors. In the words of General Blackjack Pershing—who saw firsthand the effects of weapons of mass destruction on the men under his command in World War I Europe—"We can never again afford to neglect the question of chemical and biological preparedness again."

Given the extensive history of the effective use of these weapons of mass destruction—*not only against soldiers, but against civilian populations*—it is not at all surprising that terrorist groups turn to these weapons as a weapon of first choice, not a weapon of last resort. That is why this field manual is your essential companion to help you begin to deal with this very real threat.

Chapter 4

NUCLEAR AGENTS: DETECTION, PROTECTION, TREATMENT, AND DECONTAMINATION

4.1 Introduction

The effects of nuclear weapons are qualitatively different from those of biological or chemical weapons. A nuclear detonation produces its damaging effects through blast, thermal energy, and radiation. While the most likely attack scenario from a terrorist would involve a radiological dispersal device, or dirty bomb, we will first cover nuclear blast.

Radiation includes initial radiation that directly injures humans and other life-forms, an electromagnetic pulse that directly damages a variety of electrical and electronic equipment, and residual radiation directly induced or spread by fallout that may remain at lethal levels for extended periods of time. The principal physical effects of nuclear weapons are blast, thermal radiation (heat), and nuclear radiation. These effects are dependent on the yield (or size) of the weapon expressed in kilotons (KT), the physical design of the weapon, and the method of employment. The altitude at which the weapon is detonated has a major influence on the blast, thermal, and nuclear radiation effects. Nuclear blasts are classified as air, surface, or subsurface bursts.

An airburst is a detonation in air at an altitude below 30,000 meters, but high enough that the fireball does not reach the land or water surface. The altitude is varied to obtain the desired tactical effects. Initial radiation will be a significant hazard; but there is essentially no local fallout. However, the ground immediately below the airburst may have a small area of neutron-induced radioactivity. This may impose a hazard to civilians living and passing through the area. A surface burst is a detonation in which the fireball actually touches and vaporizes the land or water surface. In this case, the area affected by blast and thermal and initial nuclear radiation will be smaller than for an airburst of comparable yield. Hiroshima and Nagasaki were technically airbursts, but at low altitude, thus creating surface burst type effects. However, in the region around ground zero, the destruction will be much greater and a crater is often produced. Additionally, all material that was within the fireball becomes fallout and will be a hazard downwind. A surface burst is the most likely type of terrorist detonation.

A subsurface burst is an explosion in which the detonation is below the surface of land or water. Cratering will always result. If the burst does not penetrate up to the surface, the only hazard is from the ground or water shock. If the burst penetrates the surface, blast, thermal, and initial nuclear radiation will be present, although less than for a surface burst of comparable yield. Local fallout will be heavy over a small area.

A high-altitude burst occurs above 30,000 meters. Radiation and physical effects are minimal at the ground level and there is no local fallout. This detonation is significant if the target is the damage of satellites with the electromagnetic pulse. Nonhardened (insufficiently protected) electronic equipment—including medical equipment—may become inoperative as with surface and even subsurface bursts. An airburst at 15,000 meters or less could be accomplished by a terrorist who managed to get a nuclear weapon aboard a commercial airliner or corporate jet. A high-altitude burst is outside the technology or capability of terrorist organizations at present—save the exception of countries such as DPRK (North Korea) who might be able to launch such a device on a crude ballistic missile.

Radiation associated with a nuclear detonation or other explosion of a nuclear device includes: (1) alpha radiation, which travels only a few centimeters and presents an internal hazard only when inhaled or ingested; (2) beta radiation, which travels a few meters in air, has limited penetrating power, and presents an external and internal hazard primarily to the skin and eyes; and (3) gamma radiation, which travels at the speed of light and can only be shieldedby heavy materials such as lead, steel, and concrete. (4) Neutron radiation, which can travel anywhere from hundreds of yards to several miles, consists of very heavy neutral particles acting as millions of tiny bullets penetrating the body and causing severe cell damage. Neutrons are unlikely to be found in a dirty bomb, but are best shielded by materials containing hydrogen, such as plastics, water, and fuel. All radiation can be deadly.

In the view of some authorities, theft of a nuclear device or building a weapon "in house" are the least-probable courses of action for a prospective nuclear terrorist. Far more likely—for a number of reasons—is the dispersal of radiological material in an effort to contaminate a target population or distinct geographical area. The material could be spread by radiological dispersal devices—that is, "dirty bombs" designed to spread radioactive material through passive (aerosol) or active (explosive) means.

Such a dispersal device may become the terrorist weapon of choice since it can be produced by anyone with access to industrial or medical radioisotopes. The extent of the area contaminated by an exploded radiological dispersal device depends on the amount of radioactive and explosive material used. Alternatively, the material could be used to contaminate food or water. This latter option is, however, considerably less likely given the huge quantities of radioactive material that would be required. The fact that most radioactive material is not soluble in water means that its use by a terrorist would be unlikely and impractical, if the purpose is to contaminate reservoirs or other municipal water supplies, because the radioactive material will settle out or be trapped in filters. Those factors, coupled with the fact that most radioactive material will present safety risks, in the form of gamma radiation exposure, to the terrorists themselves, collectively indicate the serious difficulties for any adversary attempting to store, handle, and disseminate it effectively. The major exceptions would be terrorists employing a radiological dirty bomb using plutonium and uranium—both can be handled without significant risk to the "Bomb-Builder", and yet would create mass panic if dispersed in a populated area. The actual threat is through inhalation or ingestion of the deposited material on living surfaces and foodstuffs.

Beyond the initial blast, radiological weapons kill or injure by exposing people to radioactive materials, such as cesium-137, iridium-192, iodine-131, and cobalt-60. Victims are irradiated when they get close to or touch the material, inhale it, or ingest it. With enough levels of exposure, the radiation can sicken or kill. Radiation (particularly gamma rays) damages cells in living tissue through ionization, destroying or altering some of the cell constituents essential to normal cell functions. The effects of a given device will depend on whether the exposure is "acute" (that is, brief, one time) or "chronic" (that is, an extended period).

There are a number of sources of material that could be used to fashion a non-fission nuclear device like a dirty bomb, including nuclear waste stored at a power plant, radiological medical isotopes found in many hospitals and research laboratories, and radiography sources used to inspect (essentially X-Ray) critical welds. Although spent fuel rods are sometimes mentioned as potential sources of radiological material, they are extremely radioactive, heavy, and difficult to handle, thus making them a poor choice for terrorists. A typical expended fuel rod from a commercial, test, or Navy reactor, if taken from its storage location in an unshielded manner, would create a lethal dose to the thief within as little as one to five minutes. Other sources, such as medical devices, might be much easier to steal and handle. These materials, however, generally have a lower total activity than that in a commercial reactor fuel rod bundle (although large unshielded sources are quite danger-ous). Medical and radiography devices are shielded with heavy lead "Pigs". Think of this as a small piece/nugget of highly radioactive material in a heavy lead can or cylinder. In their typical use they are remotely controlled to expose the radioac-tive material and provide the intended radiation field—thus, an intact device could be stolen and transported by a terrorist with no exposure. Presumably, terrorists could steal a medical device (either in transit or at the service facility or user location) and remove the radioactive materials. Radioactive materials are often baked into ceramic or metallic pellets. Terrorists could then crush the pellets into a powder and put the powder into a dispersal device. It could then be placed in or near a target facility or area and detonated, spreading the radiological mate-rial through the force of the explosion and in the smoke of any resulting fires. The wise terrorist, if not suicidal, would simply attach a large enough explosive incendiary device to disperse the material from within its Pig, as he/she would probably die in the act of removing the material to create a dirty bomb.

Although incapable of causing tens of thousands of instan-taneous casualties, a radiological device, in addition to possibly killing or injuring people who came in contact with it, could be used to render symbolic targets or significant areas and infra-structure uninhabitable and unusable without protective cloth-ing. A combination fertilizer truck bomb, if used together with radioactive material, for example, could not only have severely damaged one of the New York World Trade Center's towers in 1993, but might have rendered a considerable chunk of prime real estate in one of the world's financial nerve centers indefinitely unusable because of radioactive contamination. The disruption to commerce that could be caused, the attendant publicity, and the enhanced coercive power of terrorists armed with such "dirty" bombs is terrifying.

4.2   Detection of a Nuclear or Radiological Attack

Nuclear attack will probably come without warning. There will be a bright flash, enormous explosion, high winds, and a mushroom-shaped cloud indicating a nuclear attack from a true fission or fusion weapon. The first indication will be very intense light. Heat and initial radiation come with the light, and blast follows within seconds. If terrorists acquired and detonated a nuclear device, the attack would likely come without warning, and therefore initial reactions must be automatic and instinctive.

If terrorists use a radiation dispersal device in lieu of a nuclear bomb (a far more likely scenario), its detection may be delayed. In this scenario, the radioactive material is blown up using conventional explosives and is scattered across the targeted area as debris. This type of terror weapon would cause close proximity conventional medical casualties, and patients contaminated with radioactive material. While it might seem impractical to assign a high probability to terrorist attack involving radiological dispersal, given the extraordinarily poor safeguards of radioactive materials in some parts of the world, we must remain alert to this possibility.

FIGURE 4.1
Where to Find Cover Against
Blast and Thermal Effects

4.3   Immediate Actions—Impending or Imminent Nuclear or Radiological Attack

If any warning, however brief, of an impending nuclear attack is given, the most effective defense is to seek shelter from the explosion and blast. The actions taken will be highly dependent on whether a person is in a city or town or in a rural area. Regardless of where a person is, response must be immediate.

Rural Areas

If a person is in a rural area and a nuclear explosion is imminent, the back slopes of hills and mountains provide some nuclear protection. Heat and light from the fireball of a ground level nuclear blast and the initial radiation tend to be absorbed by hills and mountains. What is not absorbed often deflects upward and away from those taking shelter because of the slope of the terrain. **If this type of terrain feature is not available, seek cover in gullies, ravines, ditches, natural depressions, caves, tunnels, culverts, home basements, and behind fallen trees to reduce nuclear casualties.** Figure 4.1 illustrates the effectiveness of some of these sheltering techniques.

While a nuclear blast in a rural area is not in keeping with a likely terrorist scenario, it cannot be ignored. In a rural area or open country, you have to get something between you and the blast. Earth is a good blast protector and provides some shielding from radiation. Overhead cover with whatever is available can help protect from immediate fallout. Such a blast will make every citizen a soldier on a nuclear battlefield and the best defense is a good foxhole. Since most of us don't go about with our entrenching tools, you must seek the best shelter available, and fast. This may mean that all that you can do is stay in your car with the windows shut, and park in a protected area. In the unlikely event there is ample warning and the location of the detonation is known, your best defense is to put as much distance between you and the blast as possible. Your most likely choice for minimizing your risk is to get to the most interior part of your home or place of business (basements are good), shut all doors and windows, and put a wet towel over your nose and mouth.

Urban Areas

Though citizens should give thought to dealing with a nuclear explosion in rural areas, a far more likely scenario is that a person would encounter a nuclear explosion in an urban or suburban area. The two reasons for this are self-evident: (1) more people live in urban areas than rural ones and (2) terrorists who want to make a statement are much more likely to target a large urban area in order to maximize the number of casualties and to garner the greatest amount of publicity.

If a nuclear explosion is imminent, immediately seek shelter in a strong building. Certain types of buildings offer excellent shelter from nuclear hazards and offer the best protection when there is little time to seek shelter. Choose buildings carefully. The stronger the structure, the better the protection against blast effects. The strongest are heavily framed buildings of steel and reinforced concrete. The worst choices are the shed-type industrial buildings with light frames and long beam span. Even well-constructed frame houses are stronger than the latter. Figure 4.2 shows some typical structures that provide good protection. Keep in mind that a major cause of death in blast events is caused by the flying glass shards penetrating people

REINFORCED CONCRETE
STRUCTURE

REINFORCED
MASONRY-BRICK
HOUSE

FIGURE 4.2
Structures That Provide
Some Protection in Nuclear Blasts

near exterior windows. Characteristics to look for in selecting a
building for shelter include the following:

Pre–World War II design and construction. These have
thick, full-span floor and ceiling beams; heavy roofing
tiles; dense, reinforced walls; and, in most cases, a full
basement.

Full basements constructed of concrete or stone. Make
sure there is an exit directly to the outside as well as to the
upper floors.

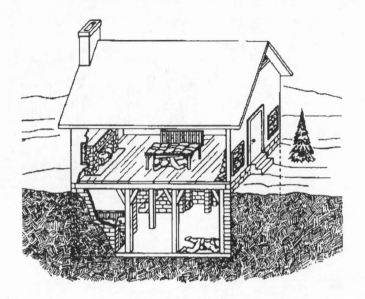

FIGURE 4.3
Where to Take Shelter in a Building
During a Nuclear Attack

Thick-walled masonry structures. A thickness of 18 inches or more is indication of solid, pre–World War II wall construction.

Buildings with the least amount of glass. Where there is glass, look for situations where windows and doors are protected by shutters.

A shielded building is best. Exterior rows of buildings in closely arranged groups shield buildings in the interior. These shielded structures suffer less blast overpressure and structural damage than exposed structures. However, debris and rubble problems and fire hazards may increase toward the center of a town.

Once inside a building, attempt to get below ground level if at all possible. The basement, because it is below ground, provides increased blast protection and much more line-of-sight radiation protection than aboveground floors. Basement parking garages provide good shelter in a nuclear attack; take cover in corners, however, not in open areas. Underground protection is increased by the surrounding earth fill. If time allows and materials are available, more protection can be gained by sandbagging a smaller shelter in the basement (such as a sturdy table) without increasing the possibility of the entire floor collapsing. Block windows with sandbags, and enhance the radiation protection and structural strength of any aboveground exterior walls by piling dirt and sandbags against the walls. **Generally speaking, you can reduce radiation by a factor of 10 in basements as compared to levels in aboveground floors.**

Your position inside the building can make a difference if sufficient time is available to properly prepare it. On floors aboveground, the center of the building offers the greatest protection from both initial and residual radiation. Below the ground, the corners of the building give the greatest protection. In either case, the dose to a prone person would be about one-half the dose to a standing person. **In short: seek shelter in an underground structure and lie in a corner.** If an underground shelter is not available, lie in the center of a shelter under a sturdy table or other furniture (Figure 4.3). Other options include lying inside a fireplace, under a stairway, or in a bathroom where the plumbing and relatively close spacing of walls might provide increased structural strength.

4.4   Immediate Actions—Actual Nuclear or Radiological Attack

It is highly likely that a nuclear or radiological attack could occur with absolutely no warning, or at best, during a period of heightened terrorist alert. In either case, mental preparation and well-honed instincts are the best defense. Though a radiological attack will not carry with it the blast and heat features of a nuclear detonation, it is important to follow similar procedures in order to provide maximum protection against radiation.

If a nuclear explosion occurs without warning, and you are in an exposed area, you should do the following:

* Immediately drop facedown; a log, rock, or any depression affords some protection.

* Close your eyes and keep them closed until the heat of the blast dissipates.

* Protect the skin from heat by putting hands and arms under or near the body.

* Remain facedown until the blast wave passes and debris stops falling.

* Stay calm, check for injuries, and prepare to help others.

Does this guidance sound familiar? Then you are from a generation that remembers the "duck and cover" atomic blast defense from the 1950s.

If a nuclear explosion occurs without warning, people in sheltered areas and buildings should also take protective actions. A blast wave can enter the shelter or building with great force, and the debris it carries can cause injuries. Lying facedown on the floor offers better protection. Attempt to avoid the violent flow of air from doors or windows or ventilation ducts. Lying near a wall is likely safer than standing away from a wall. Near a wall, reflection may increase the pressure wave, but it is better than risking being blown completely out of the building and injured by the blast.

4.5 Actions to Take After a Nuclear or Radiological Attack

Protection must not stop when an attack ends. Immediately after an attack, you must check for, or assume, radioactive contamination, and then must reduce the hazard with basic decontamination procedures. Covering your mouth with a handkerchief or other cloth, preferably a wet cloth, reduces the contaminants entering the lungs and digestive tract. This method is generally as effective as masking with the cheap masks you use when painting or spraying lawn and garden care products.

Since a radiation injury victim outside of the high neutron and gamma field near the blast does not show symptoms immediately after exposure, except for nausea and vomiting, these initial symptoms are not reliable by themselves to evaluate casualties or treat the exposed. Currently, the only available method to cheaply and quickly estimate the radiation injury to a person is with a personnel dosimeter. Without this dosimetry, many days must pass before definitive diagnosis of the secondary radiation exposure symptoms can provide an accurate estimate of radiation injury. As citizens, few of us have personal dosimeters. Medical, military, and hazardous material (HAZ-MAT) personnel in the area may be equipped with these, and their exposure may be an indication of the local level of gamma, and possibly neutron radiation exposure. Only more sophisticated dosimeters monitor for neutron dose.

4.6   Medical Defense After a Nuclear or Radiological Attack

The physiological effects of nuclear weapons are the result of exposure to the blast, thermal radiation, ionizing radiation (initial and residual) effects, or a combination of these. For smaller weapons (less than 10 KT), ionizing radiation is the primary creator of casualties requiring medical care outside the zone where the same physical damage could be caused by a conventional explosive blast, whereas for larger weapons (greater than 10 KT), the thermal, blast, and radiation impacts extend to greater distances and create larger numbers of casualties. By way of comparison, the Hiroshima blast was 15 KT, and Nagasaki was 21 KT.

The rapid compression and decompression of blast waves against the human body (called barotrauma) results in transmission of pressure waves through the tissues. Resulting damage is primarily at junctions between tissues of different densities (bone and muscle), or at the interface between tissue and airspace. Lung tissue and the gastrointestinal system (both contain air) are particularly susceptible to injury. The tissue disruptions can lead to severe hemorrhage or to an air embolism. Either can be fatal.

The thermal radiation emitted by a nuclear detonation causes burns in two ways—by direct absorption of the thermal energy through exposed surfaces (flash burns), or by the indirect action of fires in the environment (flame burns). Indirect flame burns can easily outnumber all other types of injury.

Thermal energy travels outward from the fireball in a straight line, therefore, the amount of energy available to cause flash burns decreases rapidly with distance. Close to the fireball, all objects will be incinerated. The range of 100 percent lethality will vary with yield, height of burst, weather, environment, and immediacy of treatment.

Indirect (flame) burns result from exposure to fires caused by the thermal effects in the environment, particularly from ignition of clothing, vegetation, and structures. The larger-yield weapons are more likely to cause firestorms over extensive areas. There are too many variables in the environment to predict either the incidence or severity of casualties. The respiratory system may be exposed to the effects of hot gases produced by extensive fires, and respiratory system burns cause high morbidity and high mortality rates.

The initial pulse of radiation in the optical and thermal wavelengths can cause injuries in the forms of flash blindness and retinal scarring. The brilliant flash of light produced by the nuclear detonation causes flash blindness. This flash swamps the retina, bleaching out the visual pigments and producing temporary blindness. Retinal scarring is the permanent damage from a retinal burn.

The great physical damage to the surrounding area as a result of a nuclear detonation will increase delays in medical assistance and evacuation. Quality self-aid and buddy aid will improve casualty survival rates and conserve medical resources. Prompt stabilization of victims will ensure that they can better withstand evacuation to appropriate medical treatment facilities.

Blast, thermal radiation, and nuclear radiation all cause nuclear casualties. Except for radiation casualties, nuclear casualties should be treated the same as conventional casualties. Wounds caused by blast are similar to those suffered in a severe auto accident or other serious injury. Thermal burns are treated like any other type of burn.

Local first aid cannot help radiation casualties. The best defense against radiation is to decontaminate as rapidly as possible. In short, if you survived the initial blast, take a shower or wipe down with a wet rag, put on new clothing that was stored in a closed area (e.g., drawer), and continue to avoid inhalation and ingestion of airborne and deposited material. In the early stages, this is best accomplished by keeping the wet rag over the face and mouth, and only eating items which are canned or which were stored in a sealed refrigerator or freezer. If you were lucky enough to have been in a structure not destroyed, keep the windows and doors shut, and turn off any ventilation to the outside. Unless you have totally sealed your home, you need not worry about suffocating, as there will be sufficient air exchange to provide oxygen—but what you will accomplish is a great reduction in infiltration of the residual particulate radiological materials in the outside air. For a more comprehensive understanding of radiological effects and treatment protocols for acute high-dosage radiation, consult Appendix A.

4.7   General Aspects of Decontamination

Contamination occurs when radioactive material is deposited in an area. Decontamination is the removal of radioactive particles from people, animals, or inanimate objects.

* Personal decontamination is decontamination of yourself.

* Casualty decontamination refers to the decontamination of other victims.

* Civilian decontamination usually refers to decontamination of the uninjured public.

* Mechanical decontamination involves measures to remove radioactive particulates; an example is the filtering of drinking water.

Radiological decontamination is performed in an identical manner to chemical decontamination. The main difference is timing. Chemical decontamination is an emergency. Radiological decontamination is not.

Decontamination of casualties is an enormous task. The process requires dedication of large numbers of medical or EMT/HAZMAT personnel and large amounts of time. Even with appropriate planning and training, significant resources are needed.

Victims from contaminated areas may have fallout on their skin and clothing. Although the individual will not be a significant radioactive risk to people rendering assistance, he or she may suffer radiation injury from the contamination. Removal of the contamination should be accomplished as soon as possible, and definitely before admission into a clean treatment area. In general, life threatening medical conditions take priority over radiological contamination of the victim.

Discarding of outer clothing and rapid washing of exposed skin and hair can remove 95 percent of contamination. The 0.5 percent hypochlorite solution used for chemical decontamination will also remove radiological contaminants (see Section 5.8

for preparation instructions), but even plain old soap and water work well. Care must be taken to not irritate the skin. If the skin becomes damaged, some radionuclides can be absorbed directly through the skin. Surgical irrigation solutions should be used in liberal amounts in wounds. All such solutions should be removed by suction instead of sponging and wiping. To rinse the eyes, use copious amounts of water, normal saline, or eye solutions.

Routine patient decontamination should be performed under the supervision of medical personnel. Moist cotton swabs of the nasal mucosa from both nostrils should be obtained, labeled, and sealed in separate bags. Significant decontamination will occur in the normal emergency evalua-tion of patients by careful removal and bagging of clothing.

Contaminated tourniquets should be replaced with clean ones, and the sites of the original tourniquets must be decon-taminated. Splints should be thoroughly decontaminated, but removed only by a physician. The new dressings when removed in the operating room should be placed in a plastic bag and sealed. Wounds should be covered when adjacent skin is decontaminated so skin contaminants do not enter the wound.

All casualties entering a medical unit after experiencing a radiological attack are to be considered contaminated unless there is a certificate of decontamination from the relevant civil or medical authorities. The initial management of a victim con-taminated by radiological agents is to perform all immediate life/limb-saving actions without regard to contamination. Removal of exterior clothing during the course of resuscitation will remove nearly all contamination except where the skin is exposed or the clothing is loose enough to allow contaminants to enter.

During initial decontamination in the receiving areas, bandages are removed and the wounds are flushed; the bandages are replaced only if bleeding occurs. **It is seldom possible for a living patient to be so contaminated as to pose a threat to medical providers.** Particulate matter should be removed from wounds whenever possible. Any Alpha-, Gamma-, or Beta-emitting material left in the wound will cause extensive local damage and may be absorbed and redistributed as internal contaminants. **After determination that adequate decontamination has been accomplished, the wound should again be thoroughly irrigated with saline or other physiological solutions.**

Burns should be thoroughly irrigated and cleaned with mild solutions to minimize irritation of the burned skin. Blisters should be left closed; open blisters should be irrigated and treated in accordance with appropriate burn protocols. In serious burns, radioactive contaminants are often embedded in the charred area. Because there is no circulation in the burned tissue, contaminants will remain in the layers of dead tissue. Excision of the wound is appropriate when surgically reasonable. Radioactive contaminants will be in the wound surfaces and will be removed with the tissue.

The Cold War specter of nuclear war between the superpowers conveyed the impression—still held by many—that the launch of nuclear weapons automatically meant Mutually Assured Destruction, and so there was no point in planning because there could be few survivors. While nuclear and radiological weapons in the hands of terrorists can cause terrible destruction, the amount of this material that a terrorist is likely to acquire could be quite small. Therefore, following the commonsense procedures detailed in this chapter could well make the difference in your survival should terrorists strike with these weapons.

The material presented here on nuclear blast dwells on the unthinkable. Even the event of a dirty bomb is difficult to contemplate. But the reality is that there are people out there who *are* seriously contemplating nuclear attack, and given the opportunity and circumstance, they will bring it about. We have no choice but to consider such an attack, and how we might respond during and following an attack.

Chapter 5

BIOLOGICAL AGENTS: DETECTION, PROTECTION, TREATMENT, AND DECONTAMINATION

5.1   Introduction

Biological attack is the intentional use, by an enemy, of live agents or toxins to cause death and disease among citizens, animals, and plants.
Biological agents include pathogens and toxins that produce illness and death in humans and other life-forms. Consult Appendix B for detailed information regarding specific biological agents. This appendix, adapted from military field manuals, contains information on bacterial agents, viral agents, and biological toxins. For each agent or toxin, critical information regarding signs and symptoms, diagnosis, treatment, prophylaxis and decontamination is provided in an easy-to-find format.

Biological agents pose a threat due to four primary factors: (1) small quantities of a biological agent can produce lethal or incapacitating effects over an extensive area, and can replicate in the affected hosts to cross-infect others; (2) they are relatively easily produced; (3) they are easily concealed; and (4) the variety of potential biological agents significantly complicates effective detection, protection, and defense. These factors, combined with difficult-to-detect release points, delayed onset of symptoms, and difficulties in identification with current technology, all aid terrorists who decide to use biological agents as weapons.

Medical defense against biological warfare or terrorism is unfamiliar to most military and civilian health care providers during peacetime. In the aftermath of the Gulf War in 1991, it became obvious that the threat of biological attacks against our soldiers was real. Incidents and threats of international and domestic terrorism appear to be on the rise (for example, the New York City World Trade Center bombing, the Tokyo sub-

way sarin release, the Oklahoma City federal building bombing, the Atlanta Centennial Park bombing, the 9/11 attacks on
the World Trade Center and the Pentagon, and the U.S. Senate
anthrax letters—an incident covered extensively in Tennessee
Senator Bill Frist's recent book on bio-terrorism, *When Every
Moment Counts*). In addition, numerous anthrax hoaxes
around the country have brought the issue home to civilians as
well. Other issues, including the disclosure of a sophisticated
offensive biological warfare program in the former Soviet
Union, have reinforced the need for increased training and education of health care professionals and citizens alike on how to
recognize and treat biological warfare victims.

Numerous measures to improve preparedness for and
response to biological warfare or terrorism are being developed
and implemented at local, state, and federal levels. Training
efforts have increased in both the military and civilian sectors.
The Medical Management of Chemical and Biological
Casualties course taught by the U.S. Army now trains over 500
military medical professionals each year. The Army's three-day
satellite course on the Medical Management of Biological
Casualties has reached over 40,000 medical personnel over the
last three years. But little or nothing has been done to educate
ordinary citizens like us.

Though this chapter contains a great deal of information
designed for use by medical professionals, it also contains valuable information for the general public. You will learn that
effective medical countermeasures are available against many of
the bacteria, viruses, and toxins that might be used by a terrorist as biological weapons. This education is important; it is
hoped that our physicians, nurses, and emergency medical personnel—along with the public—will develop a solid understanding of the biological threats we face and the medical
information useful in defending against these threats.

The current threat of biological warfare is serious, and the potential for devastating casualties is high for certain biological agents. Today there are at least 10 countries around the world that have offensive biological weapons programs. However, with appropriate use of medical countermeasures and common sense, many casualties can be prevented or minimized.

5.2   Distinguishing Between Natural and Intentional Disease Outbreaks

With a covert biological agent attack, the most likely first indicator of an event would be an increased number of patients showing up at hospitals with clinical features caused by the disseminated disease agent. Therefore, health care providers must assist epidemiologists to detect and respond rapidly to a biological attack. We citizens have only two courses of action available when an attack occurs: remove ourselves from the threat and protect ourselves from the threat. This may also include remaining in place so as not to move to a contaminated area. Additionally, we can all do our best to help those who must detect the attack and respond.

Many diseases caused by weaponized, or even non-weaponized, biological agents are difficult to recognize as stemming from a biological attack. The disease pattern that develops may be an important factor in differentiating between a natural and a terrorist or warfare attack. Epidemiologic clues that can potentially indicate an intentional attack are listed in Table 5.1.

Well before any such event, public health authorities must implement surveillance systems so they can recognize patterns of nonspecific syndromes that could indicate the early manifestations of a biological warfare attack. The system must be

Table 5.1

Epidemiologic Clues of a Biological Warfare or Terrorist Attack

* The presence of a large epidemic with similar symptoms, especially in a discrete population

* Many cases of unexplained diseases or deaths

* More severe symptoms than are usually expected for a specific disease or failure to respond to standard therapy

* Unusual routes of exposure for a disease, such as the inhalational route for diseases that normally occur through other exposures like exposed skin

* A disease that is unusual for a given geographic area or transmission season

* Disease normally transmitted by a means that is not present in the local area

* Multiple simultaneous or serial epidemics of different diseases in the same population

* A single case of disease by an uncommon agent (smallpox, some viral hemorrhagic fevers)

* Unusual strains or variants of organisms or diseases that resist normal treatment

* Higher attack rates in those exposed in certain areas, such as inside a building if released indoors, or lower rates in those inside a sealed building if released outside

* Disease outbreaks of the same illness occurring in noncontiguous areas

* Intelligence of a potential attack or claims by a terrorist of a release

timely, sensitive, specific, and practical. To recognize any unusual changes in disease occurrence, surveillance of existing disease activity should be ongoing, and any variation should be followed up promptly with a directed examination of the facts regarding the change. It's not easy business. Look how hard it is to keep cruise liners free from viral sickness.

It is important to remember that recognition of and preparation for a biological attack are similar to that for any disease outbreak, but the surveillance, response, and other demands on resources would likely be of an unparalleled intensity. A strong public health infrastructure is essential to prevent and control disease outbreaks, whether they are naturally occurring or otherwise. The more each citizen knows about the health care infrastructure that must deal with such a crisis, the more we can do to help these health care professionals, first responders, and other local officials.

5.3   Detection of Biological Agents

Accurate intelligence is required to develop an effective defense against biological attack. Detection of the biological aerosol prior to its release, which is highly unlikely, may allow people to don protective clothing or seek shelter. This is the best way to minimize or prevent casualties. However, interim systems for detecting biological agents are just now being fielded in limited numbers. Until reliable detectors are available in sufficient numbers, usually the first indication of a biological attack on civilians—or unprotected soldiers—will be the ill person, or perhaps the illness in a pet.

Detector systems are evolving and represent an area of intense interest within the research and development community. Several systems are now being fielded by the military and

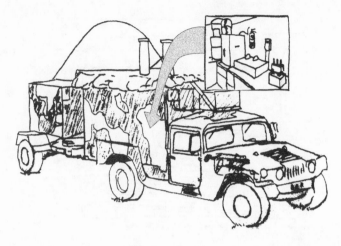

FIGURE 5.1
Military Biological Integrated Detection System (BIDS)

could be adapted for first-responder civilian use. The Biological Integrated Detection System (BIDS) (Figure 5.1) is vehicle-mounted and samples aerosol particles from environmental air, then subjects the particle sample to both genetic- and antibody-based detection schemes for selected agents. The Long-Range Biological Stand-off Detector System (LRBSDS) (Figure 5.2) is capable of providing early warning. It employs an infrared laser to detect aerosol clouds at a standoff distance up to 18 miles. An improved version is in development to extend the range to 60 miles. Of course neither of these systems is available, affordable, or practical for individual use, but may become feasible for operation by municipal or state authorities in the future.

The principal difficulty in detecting biological agent aerosols involves differentiating the artificially generated biological warfare cloud from the background of organic matter normally present in the atmosphere. Therefore, the aforementioned detection methods must be used in conjunction with intelligence, physical protection, and medical protection (vaccines

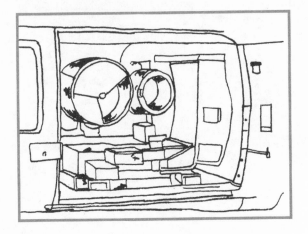

FIGURE 5.2

Military Long-Range Biological Stand-off Detector System
(LRBSDS)

and other protective measures) to provide layered primary
defenses against a biological attack.

5.4   Indications of a Biological Attack

Biological agents may be disseminated as aerosols, liquid
droplets, or dry powder. A number of biological attack indicators may signal a high probability of attack:

✳ Mysterious illness with many soldiers and civilians sick
for unknown reasons

✳ Large numbers of insects or unusual insects

✳ Large numbers of dead wild and domestic animals

✳ Mass casualties with flu-like symptoms, fever, sore

throats, skin rash, pneumonia, diarrhea, dysentery, hemorrhaging, or jaundice

✻ Mist or fog being sprayed by aircraft or aerosol generators

5.5   Immediate Actions If a Biological Attack Is Imminent

Putting on a protective mask and keeping the clothing buttoned up protects adequately against living biological agents. But an agent can gain entry through clothing using two routes: (1) openings such as button holes, zipped areas, stitching, and poor sealing at ankles, wrists, and neck; and (2) through minute pores in the fabric of clothing. Putting on protective gear, even something as simple as a raincoat, greatly increases the protection level of the individual. Toxins call for the same amount of protection as liquid chemical agents. Consider any known agent cloud as a chemical attack and take the same actions prescribed for a chemical attack (see Chapter 6).

If you are inside of a building where a biological agent has been released, follow the instructions of emergency personnel, get outside as rapidly as possible, and avoid any areas where the agent has been released. If you are outside and a biological agent has been released (for example, sprayed from an airplane), seek immediate shelter inside a nearby building. If no shelter is available, attempt to get upwind of the agent. In any event, cover all exposed skin and protect your respiratory system as soon as possible. Most residential and commercial buildings do not have efficient air filter systems, so shut down the HVAC system to avoid intake of outside air. Many buildings can be converted into temporary shelters if cracks are carefully sealed with tape or other material and a filter system with a ventilating mechanism is installed. Short of a filtered intake, natural infiltration will generally be sufficient to provide respirable levels of oxygen, while still minimizing the concentration of biological material, compared to outside atmosphere. **It must be emphasized that the most effective way to counter a biological attack is to take protective measures before an attack occurs.**

5.6   Personal Protection in a Biological Attack

Biological agents enter the body through the skin, respiratory tract, and digestive tract. The currently fielded military chemical protective equipment, which includes the M40 protective mask (Figure 5.3), battle dress overgarment (Figure 5.4), overboots (Figure 5.5), and protective gloves (Figure 5.6), will provide protection against a biological agent attack.

The M40 protective mask is available in three sizes, and when worn correctly it will protect the face, eyes, ingestion, and respiratory tract. The M40 utilizes a single screw-on filter element that involves two separate but complementary mechanisms: (1) impaction and adsorption of agent molecules onto carbon filtration media, and (2) static electrical attraction of particles initially failing to contact the filtration media available. A drinking tube on the M40 allows the wearer to drink while in a contaminated environment. Note that the wearer should disinfect the canteen and tube by wiping with a 0.5 percent hypochlorite solution before use. The military M40 mask was developed for soldiers who *live with the threat of biological warfare every day.* There are a number of masks manufactured for civilian use on the market in the $40–$50 price range, although you can pay much more.

FIGURE 5.3
M40 Protective Mask

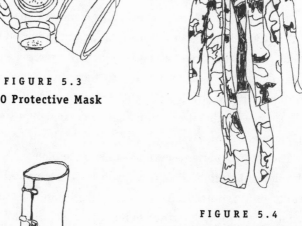

FIGURE 5.4
Military Attire: The Battle
Dress Overgarment

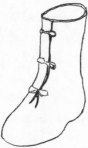

FIGURE 5.5
Military Chemical and
Biological Agent Protective
Overboots

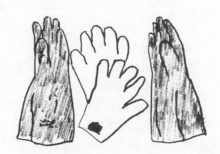

FIGURE 5.6
Military Chemical and Biological
Agent Protective Gloves

The battle dress overgarment suit comes in eight sizes and is currently available in both woodland and desert camouflage patterns. The suit may be worn for 24 continuous hours in a contaminated environment, but once contaminated, it must be replaced. The discarded garment must be incinerated or buried. Chemical protective gloves and overboots come in various sizes and are both made from Butyl rubber. The gloves and overboots must be visually inspected and decontaminated as needed after every 12 hours of exposure in a contaminated environment. While the protective equipment will protect against biological agents, it is important to note that even standard uniform clothing of good quality—or similar civilian attire— affords reasonable protection against dermal exposure of the surfaces covered.

The most important route of exposure to biological agents is through inhalation. This is why masks or even a wetted cloth can be very important. Another critical area are the eyes. In this regard, even a simple pair of swim goggles is helpful. Biological warfare agents are dispersed as aerosols by one of two basic mechanisms: point or line source dissemination. Unlike some chemical threats, aerosols of agents disseminated by line source munitions (for example, sprayed by low-flying aircraft or speedboats along the coast) do not leave a substantial hazardous environmental residue on the ground which may later affect the population (although anthrax spores may persist and could pose a hazard near the dissemination line).

Aerosol delivery systems for biological warfare agents most commonly generate invisible clouds with particles or droplets of less than 10 micrometers (μm). They can remain suspended for extensive periods, and so the major risk is inhaled particles that remain in the lungs. To a much lesser extent, particles may adhere to an individual or his clothing, thus the need for individual decontamination. The effective area contaminated varies with many factors, including wind speed, atmospheric stability class, humidity, and sunlight. In the absence of an effective real-time alarm system or direct observation of an attack, the first clue would be mass casualties fitting a clinical pattern compatible with one of the biological agents. This may occur hours or days after the attack.

Toxins are frequently as potent—or more potent—by inhalation than by any other route. Mucous membranes are also vulnerable to many biological warfare agents. Physical protection is therefore quite important and the use of full-face masks equipped with small-particle filters, like the chemical protective masks, are essential. Again, eye protection is also important.

Other routes for delivery of biological agents are thought to be less important than inhalation, but are nonetheless potentially significant. Contamination of food and water supplies, either purposefully or incidentally after an aerosol biological attack, represents a hazard for infection or intoxication by ingestion. Assurance that food and water supplies are free from contamination should be determined by appropriate preventive medicine authorities in the event of an attack.

Intact skin provides an excellent barrier for most biological agents. T-2 mycotoxins would be an exception because they attack the skin. However, mucous membranes or cuts and abrasions can allow for passage of some bacteria and toxins, and should be protected in the event of an attack.

5.7   Actions to Take After a Biological Attack

Actions to take following a biological attack include identifying victims by the symptoms they exhibit and then treating those symptoms. Early recognition of symptoms and their treatment will increase recovery time and decrease fatalities.

It is necessary to isolate people showing symptoms to disease to the degree that facilities are available. This isolation helps prevent possible spread to others if the disease is communicable. It is also necessary to limit the number of health care professionals providing care to these casualties. Treatment of live biological agent or toxin casualties requires medical assistance as soon as possible. **An indication of a live biological agent attack is a large number of police, public workers, and civilians with unexplained illness over a short period.**

A biological attack can employ a wide variety of toxins, which can be dispensed alone or with other carriers or agents. Symptoms associated with some toxins mimic other illnesses or chemical casualty symptoms. Toxin symptoms may include any of the following:

* Dizziness, mental confusion, or double or blurred vision

* Tingling of skin, numbness, paralysis, or convulsions

* Formation of rashes or blisters

* Acute coughing

* Fever, aching muscles, and fatigue

* Difficulty in swallowing

* Nausea, vomiting, and/or diarrhea

* Bleeding from body openings or blood in urine, stool, or spit

* Burning or stinging eyes

* Shock

These symptoms may appear within minutes or hours after the toxin attack. **You should decontaminate immediately after a biological attack.**

5.8   Decontamination After a Biological Attack

Contamination is the introduction of an infectious agent on a body surface, food or water, or other inanimate objects. Decontamination involves either disinfection or sterilization to reduce microorganisms to an acceptable level on contaminated articles, thus rendering them suitable for use. Disinfection is the selective reduction of undesirable microbes to a level below that required for transmission. Sterilization is the killing of all organisms.

Decontamination methods have always played an important role in the control of infectious diseases. However, we are often unable to use the most efficient means of rendering microbes harmless (for example, toxic chemical sterilization), as these methods may injure people and damage materials that are to be decontaminated. **Biological agents can be decontaminated by mechanical, chemical, and physical methods.**

Mechanical decontamination involves measures to remove but not necessarily neutralize an agent. An example is the filtering of drinking water to remove certain waterborne pathogens or, in a bio-terrorism scenario, the use of an air filter to remove aerosolized anthrax spores, or water to wash agent from the skin.

Chemical decontamination renders biological warfare agents harmless by the use of disinfectants that are usually in the form of a liquid, gas, or aerosol. Some disinfectants are harmful to humans, animals, the environment, and materials.

Physical means (heat, radiation) are other methods that can be employed for decontamination of objects.

Exposure of the skin to a suspected biological warfare aerosol should be immediately treated by soap and water. Careful washing with soap and water can remove nearly all of the agent from the skin surface. This is something we can all do. Hypochlorite solution and other disinfectants are reserved for gross contamination (that is, following the spill of solid or liquid agent from a contaminated source directly onto the skin). **Grossly contaminated skin surfaces should be washed with a 0.5 percent sodium hypochlorite solution, if available, with a contact time of 10 to 15 minutes.**

Foil packets of calcium hypochlorite are currently included in the Skin Agent Decontamination Kit (Figure 5.7) for mixing hypochlorite solutions. The 0.5 percent solution can be made by adding one 6-ounce container of calcium hypochlorite to 5 gallons of water. These solutions evaporate quickly at high temperatures, so if they are made in advance they should be stored in closed containers. Also, the chlorine solutions should be placed in distinctly marked containers because it is very difficult to tell the difference between the 5 percent chlorine solution and the 0.5 percent solution.

To mix a 0.5 percent sodium hypochlorite solution, take one part Clorox (Clorox or Clorox I; Clorox II has soap and should be avoided) and nine parts water (1:9), since standard stock Clorox is a 5.25 percent sodium hypochlorite solution. The solution is then applied with a cloth or swab. The solution should be made fresh daily.

A chlorine solution must NOT be used in (1) open body cavity wounds, as it may lead to the formation of adhesions, or (2) brain and spinal cord injuries. However, this solution may be used to wash noncavity wounds and then be removed by suction to an appropriate disposal container. Within about five minutes,

FIGURE 5.7

M291 Skin Agent Decontamination Kit

this contaminated solution will be neutralized and nonhazardous. Subsequent irrigation with saline or other surgical solutions should be performed. Prevent the chlorine solution from being sprayed into the eyes, because corneal damage may result.

Biological agents can be rendered harmless through such physical means as heat and radiation. To render agents completely harmless, sterilize with dry heat for two hours at 160 degrees centigrade (320°F). Solar ultraviolet (UV) radiation has a disinfectant effect, often in combination with drying. This is effective in certain environmental conditions but is hard to standardize for practical decontamination purposes.

The health hazards posed by environmental contamination with biological agents differ from those posed by persistent or volatile chemical agents. Aerosolized particles in the 1- to 5-μm size range will remain suspended in the atmosphere; suspended agents would eventually be made harmless by solar ultraviolet light, desiccation, and oxidation. Little, if any, environmental residues would persist. Rooms are best decontaminated with gases or liquids in aerosol form (for example, formaldehyde).

This is usually combined with surface disinfectants to ensure complete decontamination.

5.9   Ten Steps in the Management of Biological Attack Victims

Military personnel on the modern battlefield face a wide range of conventional and unconventional threats. Compared to conventional, chemical, and nuclear weapon threats, biological weapons are, perhaps, somewhat unique in their ability to cause confusion, disruption, and panic. The same is true if terrorists make your community a battlefield. It is useful to understand the factors (see Table 5.2) that account for this ability and for the difficulties in dealing with biological casualties.

In light of these somewhat unique properties of biological weapons, medical personnel will need a firm understanding of certain key elements of biological defense in order to manage effectively the consequences of a biological attack. We citizens also need this understanding in order to assist trained medical personnel as needed. In the absence of trained personnel, we will have to assist victims until the medical teams arrive. For this reason, we recommend a ten-step process to guide such evaluation and management. These medical-provider guidelines, while technical in some cases, are an excellent source of information for civilians who wish to protect themselves and their families in the event of a biological attack.

1. Maintain Your Awareness In light of the October 2001 anthrax incidents in the U.S. Capitol and DC and NJ Post Offices, every one of us should be suspicious regarding the potential employment of biological weapons or devices. This is crucial because, with many of the biological warfare diseases, early treatment is necessary if patients are to be saved. Anthrax, botulism, plague, and smallpox are readily prevented if patients are provided proper antibiotics and/or immunization promptly following exposure. Conversely, all of these diseases may prove fatal if therapy or treatment is delayed until classic symptoms develop. Unfortunately, symptoms in the early, or prodromal, phase of illness are nonspecific, making diagnosis difficult. Moreover, many potential bio-warfare diseases, such as brucellosis, Q-fever, and Venezuelan equine encephalitis, may only show themselves as little more than feverish illness.

Table 5.2

Characteristics of Biological Weapons and Warfare

✳ Potential for massive numbers of casualties

✳ Ability to produce lengthy illnesses requiring prolonged and extensive care

✳ Ability of certain agents to be contagious

✳ Lack of adequate detection systems

✳ Diminished role for self-aid and buddy aid, thereby increasing the sense of helplessness

✳ Presence of an incubation period, enabling victims to disperse widely

✳ Ability to produce nonspecific symptoms, complicating diagnosis

✳ Ability to mimic other infectious diseases, further complicating diagnosis

Without heightened awareness, it is unlikely that civilian first responders, removed from sophisticated laboratory and preventive medicine resources, will promptly arrive at a proper diagnosis and begin appropriate therapy.

2. Protect Thyself Before one can approach a potential biological attack victim, a person must first take preventive steps. This is for the same reason that, during the safety talk prior to takeoff, the airlines tell you to put your own oxygen mask in place before assisting others. These steps may involve a combination of physical, chemical, and immunologic forms of protection. The first line of physical protection typically is a protective mask. Designed primarily with chemical vapor hazards in mind, the M40 series mask certainly provides adequate protection against all inhalational biological threats. In fact, a HEPA-filter (or even a simple surgical) mask will afford some measure of protection against biological agents (although *not* against chemical agents). Immunologic protection principally involves immunization and, in the present climate, applies mainly to protection against anthrax and smallpox.

3. Assess the Patient The initial assessment is conducted before decontamination is begun and should thus be brief. Information of potential interest to the caregiver might include information about recent illnesses cropping up in the same geographic area, the sources of food and water, wind direction, immunization history, travel history, and occupational duties or location. At this point the physical assessment should concentrate on how the patient is breathing and whether he or she is coherent and has muscle control.

4. Decontaminate as Appropriate Decontamination plays an important role in the approach to chemical agent victims. The

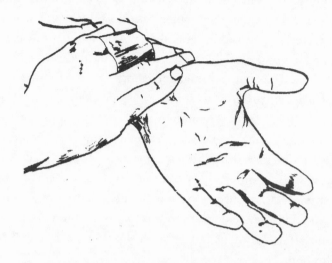

FIGURE 5.8
Skin Decontamination Using a Military M291 Kit

incubation period of biological agents, however, makes it unlikely that victims of a biological attack will show symptoms until days after an attack. At that point, the need for decontamination is minimal or nonexistent. **In those rare cases where decontamination is warranted, bathing with soap and water will usually suffice.** Certainly, standard military decontamination solutions (such as hypochlorite, if available), typically employed in cases of chemical agent contamination (Figure 5.8), would be effective against all biological agents. In fact, even 0.1 percent bleach reliably kills anthrax spores, the hardiest of biological agents. Routine use of caustic substances, especially on human skin, however, is rarely warranted following a biological attack. Information on decontamination is provided in Section 5.8.

5. Establish a Diagnosis With decontamination (where warranted) accomplished, a more thorough attempt to make a diagnosis can be carried out. Obviously few civilians will be able to make any but the simplest diagnosis (for example, difficulty breathing or skin rashes). But even the amount of expertise and support available to the clinician or medical caregiver will vary widely. A full range of laboratory capabilities may enable definitive diagnosis at many hospitals, though this is less likely in rural areas. At a minimum, every attempt should be made to obtain diagnostic specimens from representative patients for forwarding to appropriate laboratory facilities. Nasal swabs (important for culture even if the clinician is unsure which organisms to assay for), blood cultures, serum, sputum cultures, blood and urine for toxin analysis, throat swabs, and environmental samples should be considered for collection.

While awaiting laboratory confirmation, a diagnosis must be made on clinical grounds. Access to infectious disease, preventive medicine, and other specialists can assist in this process. The clinician should, at the very least, be able to make a preliminary diagnosis based on patient symptoms. Often a firm diagnosis can only be made by a trained epidemiologist, but caregivers and citizens alike can provide information to assist in that effort. Chemical and biological warfare diseases can be generally divided into those that show symptoms "immediately" with little or no incubation (principally the chemical agents) and those with a considerable delay in the onset of symptoms (principally the biological agents). Moreover, the symptoms of biological warfare diseases may resemble those of another illness. Plague, tularemia, and staphylococcal enterotoxin B disease may all show the symptoms of pneumonia. Botulism and Venezuelan equine encephalitis may exhibit peripheral and central neuromuscular findings, respectively. Table 5.3 shows a simple diagnostic matrix of rapid-onset and delayed-onset agents. Diagnosis, however, is often complicated

Table 5.3

Diagnostic Matrix of Chemical and Biological Agents

RESPIRATORY CASUALTIES

Rapid-Onset	Delayed-Onset
Nerve agents	Inhalational anthrax
Cyanide	Pneumonic plague
Mustard	Pneumonic tularemia
Lewisite	Q-fever
Phosgene	SEB inhalation
SEB inhalation	Ricin inhalation
	Mustard
	Lewisite
	Phosgene

NEUROLOGICAL CASUALTIES

Rapid-Onset	Delayed-Onset
Nerve agents	Botulism-peripheral symptoms
Cyanide	Venezuelan equine encephalitis–D central nervous system symptoms

Table 5.4

Chemical and Biological Warfare Agent Diseases Potentially Requiring Prompt Therapy

RESPIRATORY CASUALTIES

Rapid-Onset	Delayed-Onset
Cyanide	Inhalational anthrax
	Pneumonic plague
	Pneumonic tularemia

NEUROLOGICAL CASUALTIES

Rapid-Onset	Delayed-Onset
Nerve agents	Botulism

by the fact that many diseases (for example, Venezuelan equine encephalitis, Q-fever, and brucellosis) may simply look like undifferentiated feverish illnesses. Moreover, other diseases (anthrax, plague, tularemia, and smallpox) have undifferentiated feverish syndromes.

6. Render Prompt Treatment Unfortunately, it is precisely in the initial phase of many diseases that therapy is most likely to be effective. Table 5.4 shows those diseases for which prompt therapy is required, excluding those for which definitive therapy is not warranted, not available, or not critical.

Normal treatment of respiratory casualties (patients with undifferentiated feverish illnesses who might have early-stage anthrax, plague, or tularemia would be managed in a similar manner) might then be considered. Doxycycline, for example, is effective against most strains of anthrax, plague, and tularemia, as well as against Q-fever and brucellosis. Other tetracyclines and fluoroquinolones might also be considered. Keep in mind that such therapy is *not* a substitute for a careful and thorough diagnostic evaluation, when conditions permit such an evaluation. Responding to bio-terrorism is a very difficult business, even for trained medical professionals. In the absence of training or experience, ordinary citizens like us may have few treatment options other than making the victims comfortable and taking careful notes on the progress of their condition.

7. Practice Good Infection Control Standard Precautions provide adequate protection against most infectious diseases, including those potentially employed in a terrorist attack with biological agents. Anthrax, tularemia, brucellosis, glanders, Q-fever, Venezuelan equine encephalitis, and the toxin-mediated diseases are not generally contagious, and victims can be safely managed using Standard Precautions. Such precautions are certainly familiar to all clinicians and caregivers.

8. Alert the Proper Authorities In case of any kind of attack or suspected release of chemical or biological agents, local authorities should immediately be apprised of casualties and the scope of the affected area. These authorities should notify the nearest clinical laboratory and all medical facilities. This will enable laboratory and hospital personnel to take proper precautions when handling specimens and victims and will also permit the optimal use of diagnostic and treatment options.

9. Assist in the Epidemiologic Investigation All health care providers have a basic understanding of epidemiologic principles. Even under austere conditions, a rudimentary epidemiologic investigation may assist in diagnosis and in the discovery of additional biological warfare victims. Clinicians and caregivers should, at the very least, query patients about potential exposure, ill community or family members, food/water sources, presence of unusual aerosol or spray devices and wind direction that would indicate a geographical range of potential cases. Such early discovery might permit post-exposure prevention, thereby avoiding additional casualties. Preventive medicine officers, field sanitation personnel, epidemiology technicians, environmental science officers, veterinary officers, and observant citizens can all make valuable contributions to assist an epidemiologic investigation.

10. Maintain Proficiency Until recently, the threat of biological warfare has remained a theoretical one for most civilians and medical personnel. Inability to practice casualty management, however, can lead to a rapid loss of skills and knowledge among civil and medical personnel. It is imperative that these professionals maintain proficiency in responding to any terrorist situation. In part, this can be accomplished by availing oneself of several resources as listed in Appendix E. The OTSG (www.nbc-med.org) and USAMRIID (www.usamriid.army.mil) Web sites provide a wealth of information. These sites are designed primarily for health care professionals, but have valuable information for all of us. Trained and aware first responders and civilians are both invaluable in helping our communities deal with a bio-terrorism event.

Chapter 6

CHEMICAL AGENTS: DETECTION, PROTECTION, TREATMENT, AND DECONTAMINATION

6.1 Introduction

Chemical agents are used to kill, seriously injure, or incapacitate through their physiological effects. These agents have been the most prevalent of the nuclear, biological, and chemical

(NBC) agents that potential adversaries and terrorists possess and have been used in regional and internal conflicts by a number of states. This is because they are easier to produce and easier to use than biological agents. Consult Appendix C for detailed information regarding specific chemical agents. This appendix, adapted from military field manuals, contains information on pulmonary agents, blood agents, vesicants, nerve agents, and incapacitating agents. For each agent, critical information regarding signs and symptoms, detection, decontamination, and management is provided in an easy-to-find format.

The effects of chemical agents are usually immediate but can be delayed. They can be persistent or nonpersistent, and have significant physiological impact. As in the case of biological agents, a number of toxic chemical agents—including industrial chemicals—are easily produced. Though relatively large quantities of an agent are required to render an area seriously contaminated over time, small-scale selective use, especially in confined spaces such as buildings and subways, that exploits surprise can cause significant disruption and may have lethal effects.

Chemical agents, like all other substances, may exist as solids, liquids, or gases, depending on temperature and pressure. Most agents in weapons are liquids. Following detonation of a military or terrorist chemical weapon, the agent is primarily dispersed as liquid or aerosol, defined as a collection of very small solid particles or liquid droplets suspended in a gas. Thus, mustard gas and nerve gas do not become true gases, but chemicals suspended in air.

Certain chemical agents such as hydrogen cyanide, chlorine, and phosgene may be gases when encountered during warm months of the year at sea level. The nerve agents and mustard are liquids under these conditions, but they are to a certain extent volatile; that is, they volatilize or evaporate, just as water or gasoline does, to form an often-invisible vapor. A vapor is the gaseous form of a substance at a temperature lower than the boiling point of that substance at a given pressure.

The tendency of a chemical agent to evaporate depends not only on its chemical composition and on the temperature and air pressure, but also on such variables as wind velocity and the nature of the underlying surface with which the agent is in contact. Just as water evaporates less quickly than gasoline does, but more quickly than motor oil at a given temperature, pure mustard is less volatile than the nerve agent sarin, but more volatile than the nerve agent VX. However, all of these agents evaporate more readily when the temperature rises or when a strong wind is blowing.

Chemical agents are classified according to physical state, physiological action, and use. According to their physiological effects, there are nerve, blood, blister, and choking agents. The terms "persistent" and "nonpersistent" describe the time an agent stays in an area. Persistent chemical agents affect the contaminated area for an extended period of time. Conversely, nonpersistent agents affect the contaminated area for relatively short periods of time.

Chemical agents fall into five broad categories:

Pulmonary agents such as phosgene and chlorine

Blood agents, including hydrogen cyanide and cyanogen chloride

Vesicants such as mustard (H, HD), lewisite (L), and phosgene oxime (CX)

Nerve agents such as tabun (GA), sarin (GB), soman (GD), GF, and VX

Incapacitating agents such as a glycolate anticholinergic compound (BZ)

Chemical agents may enter the body by several routes. When inhaled, gases, vapors, and aerosols may be absorbed by any part of the respiratory tract. Absorption may occur through the mucosa of the nose and the mouth and/or the alveoli of the lungs. Liquid droplets and solid particles can be absorbed by the surface of the skin, eyes, and mucous membranes. Of these three, the eyes are the most critical area for absorption. Chemical agents that contaminate food and drink can be absorbed through the gastrointestinal tract. Finally, wounds or abrasions are presumed to be more susceptible to absorption than the intact skin.

The hazards from a chemical strike may last for less than an hour or for several weeks. The effects on people may be immediate. Weather factors that exert significant influence are wind, air stability, temperature, humidity, and precipitation. The best weather for direct placement of chemical weapons is calm winds with a strong, stable temperature gradient (e.g., the inversions you often hear about in reports of smog and restrictions on use of wood burning stoves). Low winds and stable or neutral conditions are most favorable for spreading an agent cloud evenly over a larger target area.

Terrorists may choose to deliver agents upwind of targets; in which case stable or neutral conditions with low to medium winds of 4–10 miles per hour are the most favorable conditions. Marked turbulence, winds above 10 miles per hour, moderate to heavy rain, and an air stability category of "unstable" result in unfavorable conditions for chemical clouds.

Although any chemical agents could be used for the purpose of causing mass casualties, it is likely that terrorists would reject most of them especially in open areas. In the case of choking agents, for example, very large amounts would be needed to inflict numerous fatalities. Blister agents, while capable of causing injury on a large scale, are very unlikely to cause death en masse. VX and other V-series nerve agents would also be unlikely candidates, because the technical challenges associated with "weaponizing" them are formidable.

Sarin, on the other hand, is highly toxic, volatile, and relatively easy to manufacture. Indeed, it was these same qualities that attracted Aum Shinrikyo's scientists to sarin and why Shoko Asahara, the group's leader, so enthusiastically supported the ambitious chemical weapons R&D program that they pursued in parallel to the cult's biological efforts. For these reasons, it is perhaps worthwhile to focus on the technological requirements needed to produce sarin, especially because it is the only chemical agent to have been employed successfully for mass-casualty purposes by a terrorist group, even though its ultimate use fell far short of the effects intended.

Although often referred to as a nerve "gas," sarin is, in fact, a liquid at any ambient temperature. When in vapor form, it is heavier than air and, as a result, will cling to floors, sink into basements, and gravitate toward low terrain. **Like all nerve agents, sarin works by interfering with the mechanisms through which one's nerves communicate with one's bodily organs. Sarin binds with a specific enzyme to destroy nerve function.** Although the effects on persons who inhale small amounts of vapor—such as occurred in the 1995 Tokyo subway attack—normally are limited to tightness in the chest, shortness of breath, and coughing, victims who inhale larger amounts soon lose consciousness, go into convulsions, and stop breathing altogether.

It has sometimes been claimed that producing sarin and other nerve agents is a relatively easy process, to the extent one authority states that "ball-point pen ink is only one chemical step removed." While sarin may be less complicated to synthesize than other nerve agents, the expertise required to produce it should not be underestimated. The safety challenges involved would, at a minimum, require skill, training, and special equipment to overcome. For this reason the level of competency needed to produce sophisticated chemical nerve agents, including sarin, is on the order of a graduate degree in organic chemistry and/or actual experience as an organic chemist—not simply knowledge of college-level chemistry, as is sometimes alleged.

Moreover, as with biological weapons, developing a means to disseminate sarin effectively is likely to prove a far greater challenge to terrorists than producing the agent itself. Although sarin's high volatility greatly simplifies its adaptability as a weapon, terrorists who may seek to cause mass casualties will need a fairly sophisticated means of spreading the agent in sufficiently large quantities over their intended target area. For wide coverage in an open area, such as a city, an airplane equipped with a suitable industrial or crop sprayer could be a satisfactory mechanism for dissemination. Alternatively, terrorists could equip a truck and drive through the target area, taking care of course, to ensure that its passengers are properly sealed off from the chemical agent. Temperature, wind speed, inversion conditions, and other meteorological factors would likely determine the effectiveness of any attack. For example, as sarin and other chemical agents are exposed to the environment, they tend to be dispersed by the wind, which requires the use of large amounts of material to ensure that a given target receives a sufficiently high dose.

In fact, the requirement to produce and disperse the large amounts of sarin or other chemical agents needed to achieve the mass-casualty levels may be the biggest disincentive for their use. A U.S. Defense Department model illustrates the problem. Releasing 22 pounds of sarin into the open air under favorable weather conditions covers about one-hundredth of a square kilometer with lethal effects. Since population densities in U.S. urban areas are typically around 5,000 people per square kilometer, such an attack would kill about 50 people. Releasing 220 pounds of sarin into the open air affects about ten times as much area and therefore would kill approximately 500 people. Releasing 2,200 pounds into the open air would cover several square kilometers, killing about 10,000 people. Thus, only in an open-air attack using amounts approaching 2,200 pounds of sarin would the effects become distinctly greater than that attainable by such traditional terrorist means as conventional explosives. One way for terrorists to overcome these problems would be to carry out an attack in an enclosed space, such as a domed stadium, office building, or subway system.

The case of the Japanese Aum Shinrikyo cult is instructive. Clearly, the cult was able to acquire the knowledge, chemicals, and equipment needed to synthesize sarin. It was an extensive research and development effort, with cost estimates as high as $30 million. Aum's 80-man program, housed in state-of-the-art facilities, was led by a Ph.D.–level scientist, and it took at least a year between the time of conception and the initial production of sarin. Nevertheless, the Tokyo subway attack, and the cult's earlier sarin attack in Matsumoto, succeeded in killing (though no less tragically) only a dozen people.

Given these impediments, a terrorist interested in harming large numbers of persons might prefer to attempt to engineer a chemical disaster using conventional means to attack an industrial plant or storage facility rather than develop and use an actual chemical weapon. In this way, significant technical and resource hurdles could be overcome, and the profile of the terrorist organization could be reduced in order to shield it from potential detection by intelligence or law enforcement agencies.

Common industrial and agricultural chemicals can be as highly toxic as bona fide chemical weapons, and, as the 1984 Bhopal, India, catastrophe demonstrated, just as devastating when unleashed on a nearby populace. In that incident, a disgruntled employee at a pesticide plant precipitated an explosion in one of the storage tanks by simply adding water to it. In the massive release of methyl isocyanine that followed, the noxious fumes affected thousands of people living near the plant. Four months later, some 1,430 persons were reported to have died as a direct result of the leak—a figure that increased to the 3,800 reported by Indian officials seven years later. A total of 11,000 persons were listed as having been disabled or harmed from exposure to the gas—in both instances, exponentially greater numbers than Aum was able to achieve in its attacks using sarin.

Yet one must consider the considerable psychological advantage that a terrorist might derive from the use of chemical weapons on U.S. soil. That alone might encourage a terrorist organization to resort to the use of chemical weapons.

6.2   Detection of Chemical Agents

Identification of chemical agents will greatly assist in the diagnosis and treatment of chemical injuries. The military detection instruments and devices described here, along with instructions for their use and procurement for civilian application, are discussed further in Chapter 7. **The following are means of detecting and identifying chemical agent contamination:**

The Military Chemical Agent Monitor (CAM) (Figure 6.1) is a handheld instrument capable of detecting, identifying, and providing relative vapor hazard readouts for G- and V-type nerve agents and H-type blister agents. The CAM uses ion mobility spectrometry to detect and identify agents within one minute of agent exposure. A weak radioactive source ionizes air drawn

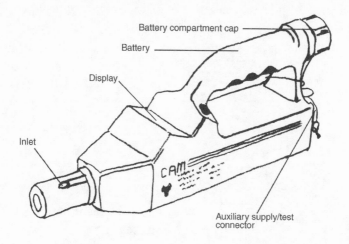

FIGURE 6.1
Chemical Agent Monitor (CAM)

into the system, and the CAM then measures the speed of the ion's movement. Agent identification is based on characteristic ion mobility and relative concentrations based on the numbers of ions detected.

The M256A1 Chemical Agent Detector Kit (Figure 6.2) is a kit that can detect and identify field concentrations of nerve agents (sarin, tabun, soman, GF, and VX), blister agents (mustard, phosgene oxime, mustard-lewisite, and lewisite), and blood agents (hydrogen cyanide and cyanogen chloride) in both vapor and liquid form in about 15–20 minutes. The kit consists of a carrying case containing 12 chemistry sets individually sealed in a plastic-laminated foil envelope, a book of M8 chemical agent detector paper, and a set of instructions. Each detector ticket has pretreated test spots and glass ampules containing chemical reagents. In use, the glass ampules are crushed to release a reagent, which runs down preformed channels to the

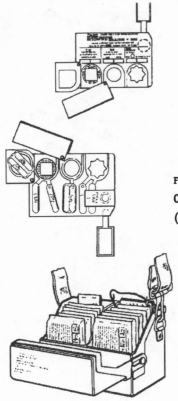

FIGURE 6.2
Chemical Agent Detector Kit
(M256A1)

appropriate test spots. The presence or absence of chemical agents is indicated through specific color changes on the test spots. The kit may be used to determine when it is safe to unmask, to locate and identify chemical hazards, and to monitor decontamination effectiveness.

The M8- and M9-type Chemical Agent Detector Paper or Tape (Figure 6.3) can be used to detect and identify liquid chemical agents or aerosols. These papers or tapes change color when exposed to liquid chemical agents or aerosols. These are convenient detectors, but often will only change color in the pres-

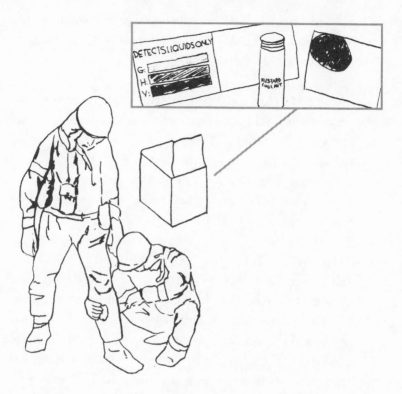

FIGURE 6.3
Chemical Agent Detector Paper or Tape (M8 and M9)

ence of a lethal dose. They are best used with masks and protective clothing.

The VGH ABC-M8 Chemical Agent Detector Paper can be used to detect and identify liquid V- and G-type nerve agents and H-type blister agents. It does not detect chemical agent vapors. The M8 paper comes in 4-by-2 1/2-inch booklets. Each booklet contains 25 sheets of detector paper that are capable of detecting G-series nerve agents (sarin, tabun, soman, and GF), V-type nerve agents, and H-type blister agents (mustard). M8 paper can identify agents through distinctive color changes from its original off-white: yellow-orange for G, blue-green for V, and red for H. M8 paper is typically used to identify unknown liquid droplets.

The M9E1 Chemical Agent Detector Paper (tape) detects the presence of liquid nerve agents (V and G) and blister agents (H/HD, HN, and L). The M9 paper does not distinguish between the types of agent involved—only that an agent or agents may be present. Neither will it detect chemical agent vapors. The M9 detector paper is rolled into 2-inch-wide by 30-feet-long rolls on a 1.25-inch-diameter core.

The Military Water Test Kit (M272) (Figure 6.4) can detect and identify hazardous levels of nerve, blister, and blood agents in treated or untreated water resources in about 20 minutes. The kit contains enough detector tubes, detector tickets, a test bottle, and prepackaged, premeasured test reagents to conduct 25 tests for each agent. Agent detection in water is indicated by the production of a specific color change in the detector tubes or in the ticket.

The Military Liquid Agent Detector (Figure 6.5) can detect droplets of GD, VX, HD, and L as well as thickened agents. It is a remote detector and transits its alarm by field wire to a central alarm unit. While primarily utilized by the military, this device may soon be fielded by emergency first responders as one method of detecting chemical agents.

Civilian chemical detection kits are now being made available to the general public. One such company, Life Safety Systems (www.lifesafetysys.com), has a number of chemical detection kits including a pocket nerve agent sensor. Most civilian defense organizations and first responders rely on the versions of the military equipment previously described.

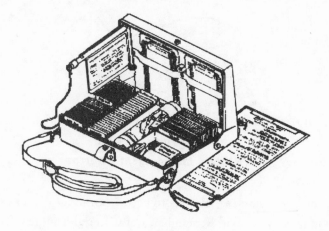

FIGURE 6.4
Military Water Test Kit (M272)

6.3   Immediate Actions If a Chemical Attack Is Imminent

If a chemical attack is imminent it is imperative that a protective mask and overgarments or chemical protective clothing be donned immediately. **Vapor exposure to the respiratory tract is the most important hazard associated with nonpersistent chemical agents. Thus, the greatest damage done by chemical agents occurs when they are inhaled because they get into the bloodstream through the blood vessels in the lungs. Therefore, if exposed to a chemical attack and protective gear is not available, attempt to seek shelter and to minimize the inhalation of the agent to the greatest extent possible.**

If you are inside of a building where a chemical agent has been released, follow the instructions of emergency personnel, get outside as rapidly as possible, and avoid any areas where the agent has been released. Use a folded handkerchief, your shirt-

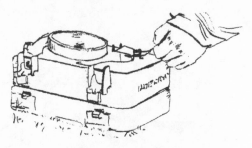

FIGURE 6.5
Military Liquid Agent Detector

sleeve, or a stack of napkins to cover your mouth and nose as you make your exit. Remember, wet material is more effective than dry material. Chemical agents are most deadly when confined to a limited area—as demonstrated by the Tokyo subway sarin attack—and therefore it is crucial to avoid confined areas where a chemical agent might be released. If you are outside and a chemical agent has been released (for example, sprayed from an airplane), seek immediate shelter inside a nearby building and call for help. In any event, cover all exposed skin and protect your respiratory system as soon as possible. **It must be emphasized that the most effective way to counter a chemical attack is to be prepared and outfitted with the right equipment before an attack occurs.**

6.4 Personal Protection in a Chemical Attack

This section details the measures by which military personnel endeavor to protect themselves in a chemical environment. This information is presented to civilian readers so that they can take similar measures to survive in the event of a terrorist attack.

Depending on the situation, military protocol may dictate the assumption of a minimum Mission-Oriented Protective Posture (MOPP) level. For civilians who do not have access to this equipment, the best personal protection can be obtained by avoiding direct contact with the chemicals to the greatest extent possible and by minimizing the extent of inhalation in the presence of agents. As one indicator of the seriousness of the chemical threat, military personnel are required to dress at MOPP 4 (consisting of wearing the protective overgarment, mask with hood, gloves, and overboots—a condition that puts extreme restrictions on mobility, and reduces effectiveness in hot weather) in all of the following situations:

* When the local alarm or command is given.

* When entering an area known to be or suspected of being contaminated with an NBC agent.

* Once chemical warfare has been initiated.

* When casualties are being received from an area where chemical agents have reportedly been used.

If individuals find themselves alone without adequate guidance, they must mask immediately and assume MOPP 4 under any of the following conditions:

* Their position is hit by artillery, mortar fire, rocket fire, or aircraft bombs and chemical agents have been used or the threat of their use is significant.

* Their position is under attack by aircraft spray.

* A suspicious odor, liquid, or solid is present.

✳ Smoke or mist of an unknown source is present or is approaching.

✳ A chemical or biological attack is suspected.

✳ Animals or birds exhibit unusual behavior and/or sudden unexplained death.

✳ Individuals have one or more of the following signs or symptoms:

✛ An unexplained sudden runny nose

✛ A feeling of choking or tightness in the chest or throat

✛ Blurring of vision and difficulty in focusing the eyes on close objects

✛ Irritation of the eyes

✛ Unexplained difficulty in breathing or increased rate of breathing

✛ Sudden feeling of depression

✛ Anxiety or restlessness

✛ Dizziness or light-headedness

✛ Slurred speech

✛ Unexplained laughter or unusual behavior

✛ Collapsing without evident cause

The mask should be worn until unmasking procedures indicate the air is free of chemical agent and the "all clear" signal is given. If vomiting occurs, lift the mask momentarily and drain it (keep your eyes closed and hold your breath), then replace, clear, and seal the mask.

Although few civilians will have the full chemical protective ensembles that military personnel use, the same procedures—maintaining a position of suspicion regarding chemical agents and protecting the respiratory tract, the eyes, and the skin to the greatest extent possible—are the most important personal protective measures that individuals can take in the event of a chemical attack.

Chemically contaminated casualties present a hazard to unprotected people. Handlers must wear their individual protective equipment when decontaminating these individuals. Casualty decontamination is conducted in an area equipped for casualty decontamination purposes. The decontamination area should be located downwind of the medical treatment facility, and contaminated clothing and equipment are placed in plastic bags and, if time allows, sealed, labeled, dated, and removed to a designated, secure repository. It is also important to conduct a brief questioning of each individual to document name, contact information, location where exposed, names of others in the exposure pathway, and general physical condition at time of decontamination. When the medical treatment facility is expected to operate in a contaminated area, collective protective shelters must be used.

Most chemical agents can poison food and water. Chemical contamination will make supplies and equipment dangerous to handle without protective equipment. Food and water packaged in sealed, airtight cans, bottles, or other impermeable containers can be decontaminated by following special procedures. Exposed foods that are known to be contaminated or suspected of being contaminated should not be consumed unless approved by medical personnel.

Police, firefighters, first responders, and medical personnel should be on the alert for the possibility of anxiety reactions among the population during chemical agent attacks. All possible steps must be taken to prevent or control such anxiety situations. Personnel in protective clothing are particularly susceptible to heat injury, and ambient temperature should be considered when determining the degree of physical activity feasible in protective clothing.

6.5   Actions to Take After a Chemical Attack

Appropriate actions following a chemical attack include identifying victims by the symptoms they exhibit and then treating those symptoms. Early recognition of symptoms and their treatment will increase recovery time and decrease fatalities. **Medical personnel should maintain a posture of suspicion, particularly about sarin and tabun, as they are the nerve agents most likely to be utilized by terrorists. Cyanide and phosgene are close seconds.** The occurrence of a number of people all exhibiting the same symptoms typically confirms that a chemical attack has taken place. **There is no vaccine for nerve agents.** Persons exposed to a nerve agent should place all of their clothing in a plastic bag and follow decontamination procedures.

6.6   Decontamination After a Chemical Attack

What follows are decontamination procedures used by military personnel. They are presented in the event that a terrorist chemical attack would render your community a chemical battlefield.

Decontamination begins with medical management, and this consists of those procedures for optimizing medical care to ensure the maximum return to duty on the battlefield. This includes triage, basic survival treatment, decontamination, emergency forward treatment, evacuation, and continuing protection of chemical agent casualties. Personal decontamination involves primarily the treatment of eyes and skin, but also involves clothing and equipment.

Following contamination of the eyes or skin with vesicants or nerve agents, decontamination must be carried out immediately. Treat the eyes first with whatever is available, and do it quickly. If possible, swab eyelids with petroleum jelly. These chemical agents are effective at very small concentrations. Within a very few minutes after exposure, decontamination is marginally effective for vesicants or nerve agents.

Decontamination consists of agent removal and/or neutralization. Decontamination after agent absorption occurs may serve little or no purpose. Service members are trained to decontaminate themselves unless they are incapacitated. For those individuals who cannot decontaminate themselves, the nearest able person should assist them as the situation permits.

Decontaminate the eyes with copious amounts of water, mild saline, or the commercially available product Artificial Tears. Decontaminate the skin with the M291 (Figure 5.7 in Chapter 5) or M258A1 Skin Decontaminating Kit (Figure 6.6).

The M291 kit contains six packets, enough to do three complete skin decontaminations. Each packet contains an applicator bag filled with decontamination powder, consisting of Ambergard WE-555 decontaminant resin. The M291 is for external use only and may be irritating to the eyes.

The M258A1 kit contains six packets: three DECON-1

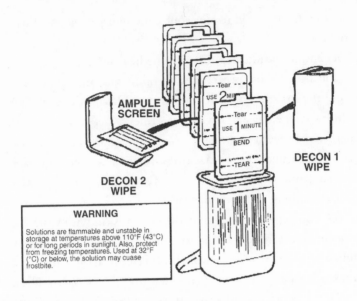

FIGURE 6.6

M258A1 Chemical Skin Decontamination Kit

packets and three DECON-2 packets. The DECON-1 packet contains a wipe prewetted with hydroxyethane 72 percent, phenol 10 percent, sodium hydroxide 5 percent, and ammonia 0.2 percent, with the remainder water. The DECON-2 packet contains a wipe impregnated with chloramine B and sealed glass ampules filled with hydroxyethane 45 percent, zinc chloride 5 percent, and the remainder water.

It is vitally important for civilians to begin the decontamination of themselves and others as soon as possible. Such timely action may have a significant effect on the degree of injury incurred by exposure to the chemical agent. There is no commercially available source or kit for chemical decontamination that I can recommend. For now, an understanding of medical and civil defense measures will help you with individual decontamination measures.

WARNING

The ingredients of the DECON-1 and DECON-2 packets of the M258A1 kit are poisonous and caustic and can permanently damage the eyes. The wipes must be kept out of the eyes, mouth, and open wounds.

Unlike biological agents, which sometimes takes days or even weeks to manifest their effects, chemical agents can act nearly instantaneously. The reaction to chemical agents must often be nearly as instantaneous. If you can successfully keep from inhaling chemical agents or getting them in your eyes, and are prepared to decontaminate your skin rapidly, you may be able to defeat the most deadly effects of chemical agents.

Chapter 7

PERSONAL DETECTION, PROTECTIVE, AND DECONTAMINATION EQUIPMENT

The most important personal detection, protective, and decontamination equipment to use in nuclear, radiological, biological, and chemical scenarios has been presented in Chapters 4, 5, and 6. Yet clearly there is more equipment that you could obtain—most of it available through suppliers who also supply directly to the military and to first responders—but the degree of extra protection afforded versus the cost and complexity of this equipment makes it unlikely that you would decide to procure a large amount of such gear.

Therefore, no attempt will be made in this chapter to provide a comprehensive listing of all possible nuclear, radiological, biological, and chemical equipment that you can buy. For those few readers who want information about this equipment, a simple search on one of the popular Internet search engines will reveal a robust list of suppliers who can and will sell you anything and everything that you could ever possibly want in this area. Appendix E lists the Web site for one of these suppliers. **Remember that these suppliers sell to both the military and civilians, but that military customers always have priority. Therefore, if you attempt to purchase this gear during times of increasing international tension, it is quite possible that the gear will be out of stock or backordered.**

This chapter presents information, in a question and answer format, regarding how the military and federal law enforcement agencies address various types of threats by using different types of equipment. In this manner, you can focus on the type of emergency or threat and learn more about the specific equipment you *might* want to obtain. **If this survival manual is to be as helpful as I hope it to be, following the commonsense procedures in Chapter 2 and Chapter 8 is far more important to your survival than "loading up" on a wide range of military equipment.** However, for those who want to know more, the following material should be useful.

Individual Protective Chemical and Biological Equipment

How do individual soldiers detect the presence of chemical agents?

While there are various new detectors for chemical and biological agents, the M8 detector paper and the M9 detector tape are both used by individuals for detection of chemical agents in the immediate environment around an individual. These are available from a number of commercial suppliers, such as Tactical Gear Command. The M256A1 and M90 are also available for chemical detection, as well as the Chemical Agent Monitor (CAM), along with the M21 Chemical Agent Point Detection System and the M42 and M21/M22 alarms.

What is the U.S. Military M8 Detector Paper?

The M8 detector paper is the only means of identifying the type of chemical agent present in liquid form on the battlefield. Each soldier is typically issued one booklet of M8 paper in the interior pocket of the gas mask carrier. The tape paper is tan in color and comes in a booklet containing 25 perforated sheets that are 2 inches by 3 inches in size and are sealed in a polyethylene envelope. A soldier encountering an unknown liquid suspected of being a chemical agent must don and check his mask within nine seconds and quickly don the attached hood and other protective gear. The soldier then removes the booklet of M8 paper from the gas mask carrier, tears a half sheet from the booklet, and if possible affixes the sheet to a stick or some other extension device. Using the stick as a handle, the soldier then blots the paper onto the unknown liquid and waits about 30 seconds for a color change. The resulting color is then compared to the colors on the inside of the front cover of the booklet to identify the type of liquid agent encountered. Yellow is G, or nonpersistent nerve agent, red is H, or blister agent, and olive or green indicates V, or persistent nerve agent. It should be noted that false positives can occur if liquid insecticides are on the surface being tested, as well as antifreeze or petroleum products.

What is the U.S. Military M9 Detector Tape?

The M9 detector tape detects the presence of liquid chemical agent, but does not identify either the specific agent or the type of agent encountered. Each soldier is typically issued one 30-foot-long and 2-inch-wide roll of M9 paper, or tape. An adhesive backing is incorporated into the tape to facilitate wrapping the paper around a sleeve or trouser leg or for attaching to equipment. Because the indicator dye in the paper is a potential carcinogen, gloves should be worn during application, and the paper should not contact the skin. The M9 tape is a dull, off-white or cream color in the absence of liquid agent, but contains an indicator chemical that, when dissolved in liquid agent, turns a reddish color. When the soldier sees the color change, the gas mask is immediately put on. The M9 paper will detect nerve agent or blister agent droplets as small as 100 microns in diameter. False positives may be seen if the paper is exposed to antifreeze, liquid insecticide, or petroleum products. The tape should not be attached to hot surfaces.

Can biological agents be detected in time to afford individual protection?

First, the accurate and efficient detection of biological agents is a complex problem. Given the chemically indistinguishable organic properties of biological agents, the methodology for detecting chemical agents is simply not useful, since each potential biological warfare agent requires a specific assay to detect and identify. Also, during the Gulf War, the U.S. military fielded only a rudimentary and developmental-stage detector system. This system could detect only two of several possible Iraqi biological warfare agent threats. In addition to the limited scope of the detector, it took between 13 to 24 hours after the attack to determine the presence and identity of the biological agent. **There was NO capability to provide any real-time or advanced warning of a biological attack.** During the Gulf War, the first likely indication of an attack was ill or dying soldiers. However, the U.S. military deployed the M31 Biological Integrated Detection System (BIDS) in 1996, which uses a multiple technology approach to detect biological agents with maximum accuracy. BIDS is a vehicle-mounted, fully integrated system that is collectively protected and is mounted on a HUMMER. Thirty-eight units were fielded, which are capable of detecting and identifying four biological agents simultaneously in less than 45 minutes. Also in 1996, the U.S. military began fielding the Interim Biological Agent Detector (IBAD), which provides shipboard detection of biological warfare agents and is capable of detecting an increase in particulate background. IBAD can detect a change in background within 15 minutes and can identify biological agents in another 30 minutes. The U.S. military also has other biological detector systems, such as the Joint Portal Shield Network Sensor System and the Hand Held Immunochromatographic Assay system, in addition to the Department of Defense Biological Sampling Kit and the Long-Range Biological Stand-Off Detector System. The M93 NBC Reconnaissance System, known as the FOX, is a high-mobility armored vehicle capable of performing NBC reconnaissance on primary, secondary, and cross-country routes throughout the battlefield. Other more advanced systems are currently in development. In the event of a domestic terrorist attack, these units could be made available to assist in the detection of biological agents.

What do U.S. federal law enforcement agencies and the U.S. military use in terms of individual personnel chemical and biological protective equipment?

The U.S. military uses a system of components for tactical and battlefield conditions that, when employed together, provides for varying degrees or stratified levels of protection against chemical and biological agents. The system is constantly being improved and updated as textile technologies progress. The system consists of a chemical protective overgarment, a chemical protective undergarment, a chemical protective helmet cover, protective overboots, protective gas masks, a field protective hood, chemical protective gloves, and various auxiliary kits such as individual decontamination kits.

What is the basic approach to individual protective clothing within the U.S. military?

In general, chemical protective clothing ensembles can be made to protect skin from chemical agents by either physical or chemical means: (1) the garment can be made of fabric that is impermeable to most substances and is basically airtight and waterproof; or (2) the garment system can be made of a fabric that is permeable to most molecules, but chemically alters or physically removes chemical agents before they reach the skin. With regard to the first method, the toxic substances are totally excluded because the agent is physically prevented from penetrating the garment material. The problem with this method is that excess heat and moisture build up since the garment is airtight, and natural cooling through perspiration is simply not possible because the water vapor cannot pass through the outer garment. Indeed, heat injury can occur using this approach, particularly when considering soldiers who are expected to carry out their duties while on the move, along with a variety of other tactical tasks. Accordingly, most military and military-related protective overgarments depend on the garment fabric's ability to adsorb the threat agent while allowing for some level of air permeability and cooling capability. Though specific garment designs vary, the basic approach to this requirement is through the use of activated charcoal in different forms. It should be noted that special equipment, not related to the stratified approach, is used for applications when higher concentrations are expected, such as for decontamination crews or stockpile monitoring.

What are the different levels of protection currently used by the U.S. military for chemical and biological protection in battlefield conditions?

Placing a soldier into full chemical protective equipment,

including mask, overgarment, gloves, and boots, is a decision that involves not only the necessary protection needed to address the threat, but also the added heat stress and potential for dehydration and decrease in force effectiveness. The physical burden of a full protective suit with the various pieces of protective gear can add about 9 to 14 pounds to a normal load. As a result, this added weight, combined with heat stress, possible dehydration, and physical exertion, can cause significant impairment to any mission. Because of these factors, the completeness of protection is stratified according to the anticipated magnitude of biological or chemical threat. This U.S. military doctrine is referred to as Mission-Oriented Protective Posture, or MOPP as we discussed in Section 6.4. There are seven basic MOPP levels, each with its own capability to afford protection and allow for physical and tactical movement.

What is the U.S. military JSLIST chemical protective undergarment?

The Joint Services Lightweight Integrated Suit Technology (JSLIST) is one of the latest additions to the military's arsenal of chemical/biological protective equipment. The chemical protective undergarment, or CPU, is a protective undergarment consisting of trousers, shirt, booties, a balaclava, and protective gloves that are all worn underneath the battle dress uniform, flight suit, or other outer military garment. The fabric contains the protection: polymerically encapsulated activated carbon as an integral component of the material fiber. Accordingly, the material does not "dust" or lose carbon to the user or the environment, thereby affording a shelf life of about 12 years. The CPU provides continuous proactive protection of chemical/biological vapors and aerosols and can be laundered for 10 times. A unique advantage of the CPU is that the protection is masked since the system is worn underneath the outer clothing.

Accordingly, critical pieces of gear, such as weapons and other personal gear, remain available for use and are unencumbered by the protective garment.

What is the U.S. military JSLIST chemical protective overgarment?

The U.S. military is continuing the acquisition of JSLIST suits as a replacement for the battle dress overgarment (BDO) and other chemical suits. The BDO will continue to be used until the end of its service life. In fact, the BDO is reaching its maximum extended shelf life limit of 14 years and there are no companies still producing it. The newest chemical protective overgarment (CPO) is the JSLIST overgarment suit, which contains carbon beads for chemical adsorption of chemical agents and provides vapor, aerosol, and liquid/splash protection. Flight suit versions are also available. The JSLIST overgarment is a universal, lightweight, two-piece, front-opening garment that can be worn as an overgarment or as a primary uniform over personal underwear. It has an integral hood, bellows-type pockets, high-waist trousers, adjustable suspenders, adjustable waistband, and waist-length jacket that enhances system comfort and is designed for use up to 45 days of continuous wear, providing 24-hour chemical protection.

For domestic law enforcement and first responder chemical/biological overgarment applications, the Federal Bureau of Investigation wears the LANX chemical protective overgarment. This innovative system is based on polymerically encapsulated activated carbon that is integral to the fiber and fabric material; the outer shell material is specially treated 50 percent cotton/50 percent nylon in a rip-stop fabric and comes in a variety of camouflage patterns and solid colors.

What types of protective gas masks are used by the U.S. military?

Several models of protective masks were used by the military in the Gulf War, and some of the same designs and models are in use today. All of the masks protect the face and airways from airborne contamination by all known chemical or biological agents, as well as radioactive dust. In the Gulf War, most U.S. troops in dismounted ground operations were issued the M17 series mask; others were issued the newly developed M40 protective mask. Both the M40 and the M17 have similar basic functions and levels of protection, but the M40 is more comfortable with improved convenience and voice transmission characteristics. They both include a binocular lens system, elastic head harness, voicemitters for making oneself heard by others, and filters to trap NBC agents. The M40 and the M42 series masks are undergoing the final stages of field testing to replace the M17, M9, and M25 series counterparts. The new versions of the 40 series masks offer increased protection, improved fit and comfort, ease of filter change, better compatibility with weapon sights, and a Butyl rubber second skin. The Army, in cooperation with the Marine Corps, recently completed a product improvement program for the M40 series mask that allows ground soldier-to-aircrew communication.

What is the M95 gas mask?

The M95 gas mask is a respirator that provides the highest levels of efficiency and comfort in modern NBC protection and meets the most critical hazards and stresses encountered in combat or law enforcement situations. The M95 filter cartridge can be mounted on the left or right using the standard 40 mm thread interface. The M95 mask has a 20-year shelf life, and the filter has a 10-year shelf life. This mask also features a drinking device, along with a six-point head harness for maximum comfort.

What is the Millennium® Chemical-Biological Mask?

The Millennium® Chemical-Biological Mask is a Hycar version of the reliable U.S. military MCU-2/P mask, used by the U.S. Air Force, which combines high performance, customized fit, comfort, and cost efficiency. This advanced mask is effective against biological agents and GA, GB, GD, VX, mustard, and lewisite chemical warfare agents. The mask features a one-piece polyurethane lens with a wide field of vision and has a dual-canister mount using the NATO standard 40 mm thread, which allows weapon sighting from either shoulder. A drinking tube provides a standardized connection for fluid ingestion in contaminated atmospheres. The six-point head harness is fully elastic and promotes easy on/off with no hair pulling. An internal nose cup with two check valves deflects air from the lens to reduce fogging.

What is the Advantage® 1000 Chemical Biological Agent-Riot Control Agent Gas Mask?

The Advantage® 1000 Gas Mask is a clean and simplified gas mask with a Hycar facepiece that is based on a proven facepiece design used by the U.S. Armed Forces. This gas mask was featured in an article in *Newsweek* for its effectiveness in protecting against toxic agents. This mask is about 40 percent lighter than conventional full-face respirators and provides high performance, a customized fit, economy, and user acceptance. Similar to the Millennium® mask, this mask also features a flexible one-piece polyurethane lens with a wide field of vision and dual-canister bayonet mounts that facilitates weapon sighting from either shoulder. The mask is fully elastic, and has a six-point head harness that promotes easy on and off and user-friendly adjustment. A standard nose cup reduces fogging and this mask also comes with a standard mechanical speaking diaphragm, or a communications system can be added. This gas mask is available from Tactical Gear Command.

Can you drink water while wearing a gas mask?

It depends on your gas mask type. The M1 chemical canteen top and tube assembly will allow users of the MCU-2A/P, M40, and M95 masks to drink using a traditional canteen fitted with the special M1 cap. Most recently, the U.S. Army approved a 2-liter reservoir for use with these same masks. This is a Camelback-type system worn on the back and provides a convenient extended source of hydration while on the move.

What is the Field Protective Hood?

The hood attaches to and is donned with the mask. It protects the head and neck from chemical agents, biological agents, and nuclear contaminants. The current JSLIST and LANX CPO parka incorporates a hood for this purpose.

What types of chemical protective gloves are available and used by the U.S. military and federal law enforcement?

The glove set includes outer gloves made of impermeable Butyl rubber and inner gloves made of thin cotton to absorb moisture. The outer gloves come in three thicknesses:

* The 7-mil gloves are used by medical personnel, computer operators, electronic repair and service personnel, and others who need high touch sensitivity and who normally will not expose the gloves to harsh treatment.

* The 14-mil gloves are used by aviators, vehicle

mechanics, and weapon crews needing some touch sensitivity but who also are unlikely to give the gloves harsh treatment.

✱   The 25-mil gloves are used by troops who perform close combat and other heavy labor.

What types of boots are used by the U.S. military for chemical and biological protection?

During the Gulf War, vinyl-based materials were used in the black vinyl overboot (BVO) and the green vinyl overboot (GVO). These boots, when worn over combat boots, provide an impermeable protective barrier against chemical, radiological, and biological hazards, as well as rain and snow. Current issue boots are called the multipurpose overboot (MULO) and have been adopted by all four military services. The MULO is a 60-day boot that provides 24 hours of chemical protection and features increased traction, improved durability, resistance to petroleum, oil, lubricant and flame, and better chemical protection than the BVO and GVO. Another boot, the toxicological agent protective boot, also called the TAP boot, provides 8 hours of protection from toxicological agents; the TAP boot is an actual boot, not an overboot, and is black with a yellow toe.

What types of decontamination kits are used by the U.S. military?

Personnel decontamination is performed to reduce the level of contamination so it is no longer a hazard to the individual. It consists of removal of clothing and decontamination of the skin. Personnel decon kits are used to remove the gross contamination. Several different types of decontamination units are used by the U.S. military. The M291 is a personnel-individual decontamination kit for the skin.

What is the U.S. military M291 Skin Decontamination Kit?

The M291 kit is a soft package consisting of two flexible pockets, each of which contains three decontamination packets. Each of the packets contains a black resin that is both reactive and adsorbent. The decontamination pad is made from a nonwoven fiberfill that is impregnated with a dry resin mixture. The decontamination is accomplished by merely opening the packet and scrubbing the skin surface with the applicator pad until an even coating of the resin is achieved. Use normal precautions to keep the powder from wounds, the eyes, and the mouth. Open wounds can be decontaminated with water, saline, or dilute hypochlorite solutions.

What is the U.S. military M272 Chemical Agent Water Testing Kit?

The M272 chemical agent kit is designed to detect and identify, via colorimetric reaction, hazardous levels of nerve agents, mustard, lewisite, and cyanide in treated or untreated water. A full kit contains enough supplies to perform 25 tests for each agent. About 20 minutes is required to perform all four tests. All bodily contact should be avoided with the kit chemicals, because some can be very harmful and should only be handled while wearing protective gloves and equipment.

What type of individual protective equipment system is used by U.S. Federal Emergency Management Agency personnel and first responders in addressing weapons of mass destruction incidents?

The U.S. Attorney General has commissioned the Interagency Board (IAB) for equipment standardization and interoperability. This board has divided personal protective levels into three categories, based on the degree of protection afforded. During weapons of mass destruction response operations, the Incident Commander determines the appropriate level of personal protective equipment. These are:

* Level A is a fully encapsulated, liquid and vapor protective ensemble designed for the highest level of skin, respiratory, and eye protection.

* Level B provides for liquid splash resistance used with the highest level of respiratory protection.

* Level C is a liquid splash-resistant ensemble with the same level of skin protection as Level B but used when the concentrations and types of airborne substances are known and the criteria for using air-purifying respirators are met.

These are only a few of the questions and answers concerning nuclear, biological, and chemical protective equipment. Once again, having a plan and exercising the plan are more important than collecting a stockpile of gear. If you have to have only one piece of gear, I recommend a mask. Your airway and your eyes are the most vulnerable avenues for the broadest range of these agents.

Chapter 8

LIVING WITH NUCLEAR, BIOLOGICAL, AND CHEMICAL TERRORISM

If I've done my job properly in compiling and editing this book, it has convinced you that the threat of nuclear, radiological, biological, or chemical terrorism is real and it is growing. I hope you now know that there are effective steps that *you can take* to limit your vulnerability and help prevent you and your family from becoming victims when, *not if,* there is a nuclear, biological, or chemical terrorist attack.

However, even following the commonsense procedures in this book, you will likely still be forced to live with this threat for the foreseeable future. How should you and your family deal with this? Do you need to be on high alert for the rest of your lives? Could you deal with the stress that accompanies such moment-to-moment vigilance? Do you need to be constantly at this alert level?

The short answer is no. There is a thin line between prudence and panic. The challenge in responding to the threat of a terrorist attack using of weapons of mass destruction is to craft defense capabilities that are both cost-effective and appropriate. But these measures need to be adaptive enough to respond as effectively as possible in a wide range of circumstances or scenarios.

The consequences of a terrorist attack involving nuclear or radiological material or a chemical or biological agent will be devastating. You cannot assume that it will happen in some other community or city, and not your own. The challenge, therefore, is to avoid reacting too strongly and still prepare adequately for a threat that, while uncertain, could nonetheless profoundly affect you and your family.

A critical step in this process is to reconsider the "worst-case-scenario" threat assessment that has dominated domestic planning and preparedness for potential acts of terrorism with weapons of mass destruction. The narrow focus on lower-probability/higher-consequence threats, which pose limitless vulnerabilities, does not reflect the realities of terrorist behavior and operations. Nor does preparing for these higher-consequence threats automatically solve all of the higher-probability/lower-consequence scenarios.

What does this mean for you? Our national leadership, from the President, to the Congress, to the Department of Defense, to the Department of Homeland Security, to the Federal Emergency Management Agency, to your state and local disaster authorities, is focused on the "big picture"—everything from working internationally and domestically to deter terrorism before it starts, to hunting down terrorists before they strike. The leadership at all levels is responsible for developing macro-scale disaster-preparedness plans. While you want to be a well-informed and prepared citizen, you have virtually no influence in any of these issues. There is little—other than being the eyes and ears of law enforcement and on the lookout for suspicious people or activities—that you can do to deal with these macro-scale issues.

What you can do? Two things.

The first is to not become paralyzed by fear. Citizens in other nations, such as Northern Ireland, Lebanon, and Israel, have been living with the daily threat of terrorism for decades and have managed to retain a balance in their lives. There is no reason why we can't do that here. Go through your daily activities with a greater sense of purpose based on the knowledge that your nation is leading the international fight against terrorism.

The second is to *actually put into practice the recommendations in this book.* We all lead busy lives. It takes time to develop a family emergency action plan or to build a family emergency supply kit, or to actually go through the drill of evacuating your house in an emergency. But as the Nike ad famously says: "Just do it." Once you have familiarized yourself with the appropriate procedures and *practiced them* you will have dampened many of your fears and anxieties. Why? Because you have taken a crucial step in protecting yourself and your family from terrorist actions. If knowledge is the antidote for fear, action is the precursor to peace of mind.

Which brings us full circle. At the outset I told you that this book gathered and clearly presented information from military manuals—from the experts. Our military has gained the upper hand on those who would use weapons of mass destruction because it has studied the problem and developed effective procedures and defenses. I have drawn from this information to help build this text—your nuclear, biological, and chemical warfare field manual. But the U.S. military has taken another crucial step—*they have trained so that they can respond to the threat of weapons of mass destruction.* You now have the training plan—this manual. Now it's up to you.

Appendix A

NUCLEAR AGENTS
AND THEIR
EFFECTS

A.1   Medical Defense

A.2   Acute High-Dose Radiation

A.3   Management Protocol for Acute Radiation
      Syndrome

A.1   Medical Defense

Introduction

Medical defense against nuclear and radiological warfare is one of the least emphasized segments of modern medical education. Fifty plus years of nuclear-doomsday predictions have made any realistic preparation for radiation casualty management an untenable political consideration. The end of the Cold War had dramatically reduced the likelihood of strategic nuclear weapons use and thermonuclear war.

Unfortunately, the proliferation of nuclear material and technology has made the acquisition and adversarial use of ionizing radiation weapons more probable than ever. In the modern era, military personnel and their nation's population will expect that a full range of medical treatment will be employed to decrease the morbidity and mortality from the use of these weapons. Fortunately, the treatment of radiation casualties is both effective and practical.

Prior to 1945, ionizing radiation was deemed nearly innocuous and often believed to be beneficial. Individual exposures to low-level radiation commonly occurred through the 1950s and early 1960s from cosmetics, luminous paints, bright wrist watch and alarm clock radium dials, medical-dental X-ray machines, and shoe-fitting apparatus in retail stores. The physical destruction caused by the nuclear explosions above Hiroshima and Nagasaki and the Civil Defense programs of the 1950s and 1960s changed that perception.

Since that time, popular conceptions and misconceptions have permeated both attitudes and political doctrine. The significance of radiological accidents at Chernobyl in the Ukraine and Goiania in Brazil are models for the response to the use of radiological weapons. To date, radiological warfare has been limited to demonstration events such as those by the Chechens in Moscow and threats by certain deposed third-world leaders.

As U.S. forces deploy to areas devastated by civil war and factional strife, unmarked radioactive material will be encountered in waste dumps, factories, abandoned medical clinics, and nuclear fuel facilities. For example, Peace-Keeping forces in the Balkans have encountered huge sources left behind from mining operations by the FSU: subsistence level peasants dumped the radioactive sources in the forest and sold the lead containers for cash. Alternatively, terrorists could use this material against citizens in any nation. Medical providers must be prepared to adequately treat injuries complicated by ionizing radiation exposure and radioactive contamination. To that end, the theory and treatment of radiological casualties are taught to both military and civilian personnel in the Medical Effects of Ionizing Radiation Course offered by the Armed Forces Radiobiology Research Institute at Bethesda, Maryland.

Radiation Threat Scenarios

A radiation dispersal device (RDD) is any device that causes the purposeful dissemination of radioactive material across an area without nuclear detonation. Such a weapon can be easily developed and used by any combatant with conventional weapons and access to radionuclides. The material dispersed can originate from any location that uses radioactive sources, such as a nuclear waste processor, a nuclear power plant, a university research facility, a medical radiotherapy clinic, an industrial complex, or pirated weapons-grade nuclear material (e.g., enriched uranium or separated Plutonium-239). The radioactive source is blown up using conventional explosives and is scattered across the targeted area as aerosol or particulate debris.

This type of weapon would cause conventional casualties to become contaminated with radionuclides and would complicate medical evacuation within the contaminated area. It would function as either a terror weapon or terrain-denial mechanism. Many materials used in military ordnance, equipment, and supplies contain radioactive components. U.S. forces may be operating in a theater that has nuclear reactors that were not designed to U.S. specifications and are without containment vessels. These reactors may be lucrative enemy artillery or bombing targets.

Significant amounts of radioactive material may be deposited on surfaces after the use of any nuclear weapon or RDD, destruction of a nuclear reactor, a nuclear accident, or improper nuclear waste disposal. Military operations in these contaminated areas could result in military personnel receiving sufficient radiation exposure or particulate contamination to warrant medical evaluation and remediation.

Depleted uranium munitions on the battlefield do not cause a significant radiation hazard, although if vaporized and inhaled, they do pose the risk of heavy-metal toxicity to the kidneys. For the layman, it might be interesting to understand that "depleted uranium" is the uranium left over after the 1 percent or less of the highly fissionable part of the ore has been extracted for use in weapons and reactor fuel—think of it like a metal just 20 percent or so heavier than lead—that is why it makes good bullets and armor plating! Materials such as industrial radiography units, damaged medical radiotherapy units, and old reactor fuel rods can be responsible for significant local radiation hazards.

Many nations have, and more may soon have, the capability of constructing nuclear weapons. The primary limitation is the availability of weapons-grade fuel. Combatants with a limited stockpile of nuclear weapons or the capability of constructing improvised nuclear devices might use them either as desperation measures or for shock value against troop concentrations, political targets, or centers of mass population. Small-yield tactical nuclear weapons might also be used in special situations. A small, functioning tactical nuclear weapon is a terrorist's dream come true.

Nuclear weapons might also be employed as a response to either the use or threat of use of any weapon of mass destruction. Large numbers of casualties with combined injuries would be generated from the periphery of the immediately lethal zone. Advanced medical care would only be available outside the area of immediate destruction. Consequently, primary management importance would be placed on evacuating casualties to a multiplicity of available medical centers throughout the United States.

Types of Ionizing Radiation

Alpha particles are massive, charged particles (four times the mass of a neutron). Because of their size, alpha particles cannot travel far and are fully stopped by the dead layers of the skin or a uniform. Alpha particles are a negligible external hazard, but when they are emitted from an internalized (e.g., lungs and GI tract) radionuclide source, they can cause significant local cellular damage in the region immediately adjacent to their physical location.

Beta particles are very light, charged particles that are found primarily in fallout radiation. These particles can travel a short distance in tissue; if large quantities are involved, they can produce damage to the basal stratum of the skin. The lesion produced, a "beta burn," can appear similar to a thermal burn. This in fact was the worst physical damage caused by radiation to any workers involved in the Three Mile Island reactor event, and then it was minor in medical terms. Probably the most sensitive human organ to Beta is the lens of the eye. Safety glasses can prevent most of the potential exposure from external Beta sources.

Gamma rays, emitted during a nuclear detonation and in fallout, are uncharged radiation similar to X-rays. They are highly energetic and pass through matter easily. Because of its high penetrability, gamma radiation can result in whole-body penetrating exposure.

Neutrons, like gamma rays, are uncharged, and are only emitted during the nuclear detonation, and are not a fallout hazard. However, neutrons have significant mass and interact with the nuclei of atoms, severely disrupting atomic structures. Compared to gamma rays, they can cause 20 times more damage to tissue. Neutron radiation, although potentially present in very minute amounts from uranium and plutonium, is not a concern in dirty bombs.

When radiation interacts with atoms, energy is deposited, resulting in ionization (electron excitation). This ionization may damage certain critical molecules or structures in a cell. Two modes of action in the cell are direct and indirect action. The radiation may directly hit a particularly sensitive atom or molecule in the cell. The damage from this is irreparable; the cell either dies or is caused to malfunction.

The radiation can also damage a cell indirectly by interacting with water molecules in the body. The energy deposited in the water leads to the creation of unstable, toxic hyperoxide molecules; these then damage sensitive molecules and afflict subcellular structures.

Units of Radiation

The *radiation absorbed dose* (rad) is a measure of the energy deposited in matter by ionizing radiation. This terminology is being replaced by the International System skin dose unit for radiation absorbed dose, the gray (Gy): 1 Gy = 100 rad; 10 milligray (mGy) = 1 rad; 1 rad = 1 centigray (cGy). The dose in gray is a measure of absorbed dose in any material. The unit for dose, gray, is not restricted to any specific radiation, but can be used for all forms of ionizing radiation. Dose means the total amount of energy absorbed per gram of tissue. The exposure could be single or multiple and either short or long in duration.

Dose rate is the dose of radiation per unit of time.

Free-in-air dose refers to the radiation measured in air at a certain point. Free-in-air dose is easy to measure with current field instruments, and more meaningful doses, such as midline tissue dose or dose to the blood-forming organs, may be estimated by approximation. Military tactical dosimeters measure the free-in-air doses.

Different radiation types have more effects as their energy is absorbed in tissue. The difference is adjusted by use of the quality factor (QF). The dose in rads times the QF yields the rem, or *radiation equivalent man*. The international unit for this radiation equivalency is the sievert (Sv) and is appropriately utilized when estimating long-term risks of radiation injury. Since the QF for X-ray or gamma radiation = 1, then for pure gamma radiation:

$$100 \text{ rad} = 100 \text{ cGy} = 1{,}000 \text{ mGy} = 1 \text{ Gy} = 1 \text{ Sv} = 100 \text{ rem}$$

NUCLEAR DETONATION AND OTHER HIGH-DOSE RADIATION SITUATIONS

A.2   Acute High-Dose Radiation

Acute high-dose radiation occurs in three principal tactical situations:

* A nuclear detonation will result in extremely high dose rates from radiation during the initial 60 seconds (prompt radiation) and from the fission products present in the fallout area relatively close to ground zero.

* A second situation would occur when high-grade nuclear material is allowed to form a critical mass ("criticality"). The subsequent nuclear reaction then releases large amounts of gamma and neutron radiation without a nuclear explosion.

* A radiation dispersal device made from highly radio-active material such as Cobalt-60, Iridium-192, or

Cesium-137 could also produce a dose high enough to cause acute injury.

The two most significant radiosensitive organ systems in the body are the hematopoietic and the gastrointestinal (GI) systems. The relative sensitivity of an organ to direct radiation injury depends upon its component tissue sensitivities. Cellular effects of radiation, whether due to direct or indirect damage, are basically the same for the different kinds and doses of radiation.

The simplest effect is cell death. With this effect, the cell is no longer present to reproduce and perform its primary function.

Changes in cellular function can occur at lower radiation doses than those that cause cell death. Changes can include delays in phases of the mitotic cycle, disrupted cell growth, permeability changes, and changes in motility. In general, actively dividing cells are most sensitive to radiation. Radiosensitivity also tends to vary inversely with the degree of differentiation of the cell. This is the reason why limits to radiation exposure in the nuclear industry are severely restricted for pregnant females, as the fetus is in a state of very rapid cell division over the 9-month gestation period.

The severe radiation sickness resulting from external irradiation and its consequent organ effects is a primary medical concern. When appropriate medical care is *not* provided, the median lethal dose of radiation, the $LD_{50/60}$ (that which will kill 50 percent of the exposed persons within a period of 60 days), is estimated to be 3.5 Gy.

Recovery of a particular cell system is possible if a sufficient fraction of a given stem cell population remains after radiation injury. Although complete recovery may appear to occur, late somatic effects may have a higher probability of occurrence because of the radiation damage.

Modern medical care dramatically improves the survivability of radiation injury. Nearly all radiation casualties have a treatable injury if medical care can be made available to them. Casualties with unsurvivable irradiation are usually immediately killed or severely injured by the blast and thermal effects of a detonation. Unfortunately, significant doses of radiation below the level necessary to cause symptoms alter the body's immune response and sensitize the person to the effects of both biological and chemical weapons.

Effects on Bone-Marrow Cell Kinetics

The bone marrow contains three cell renewal systems: the erythropoietic (red cell), the myelopoietic (white cell), and the thromobopoietic (platelet). A single stem cell type gives rise to these three cell lines in the bone marrow, but their time cycles, cellular distribution patterns, and post-irradiation responses are quite different.

The erythropoietic system is responsible for the production of mature erythrocytes (red cells). This system has a marked propensity for regeneration following irradiation. After sublethal exposures, marrow erythropoiesis normally recovers slightly earlier than myelopoiesis and thrombopoiesis and occasionally overshoots the baseline level before levels at or near normal are reached. Although anemia may be evident in the later stages of the bone-marrow syndrome, it should not be considered a survival-limiting factor.

The function of the myelopoietic cell renewal system is mainly to produce mature granulocytes, that is, neutrophils, eosinophils, and basophils, for the circulating blood. Neutrophils are the most important cell type in this cell line because of their role in combating infection. The most radiosensitive of these cells are the rapidly proliferating ones.

The mature circulating neutrophil normally requires 3 to 7 days to form from its stem cell precursor stage in the bone marrow.

Mature granulocytes are available upon demand from venous, splenic, and bone-marrow pools. These pools are normally depleted soon after radiation-induced bone-marrow injury. Because of the rapid turnover in the granulocyte cell renewal system (approximately 8-day cellular life cycle), evidence of radiation damage to marrow myelopoiesis occurs in the peripheral blood within 2 to 4 days after whole-body irradiation.

Recovery of myelopoiesis lags slightly behind erythropoiesis and is accompanied by rapid increases in numbers of differentiating and dividing forms in the marrow. Prompt recovery is occasionally manifested and is indicated by increased numbers of band cells in the peripheral blood.

Platelets are produced by megakaryocytes in the bone marrow. Both platelets and mature megakaryocytes are relatively radioresistant; however, the stem cells and immature stages are very radiosensitive. The transit time through the megakaryocyte proliferating compartment in humans ranges from 4 to 10 days. Platelets have a lifespan of 8 to 9 days. Platelet depression is influenced by the normal turnover kinetics of cells within the maturing and functioning compartments.

Thrombocytopenia is reached by 3 to 4 weeks after midlethal-range doses and occurs from the killing of stem cells and immature megakaryocyte stages, with subsequent maturational depletion of functional megakaryocytes. Regeneration of thrombocytopoiesis after sublethal irradiation normally lags behind both erythropoiesis and myelopoiesis.

Now after all of these technical discussions, the layman should understand that bone marrow transplants performed by U.S. doctors to the most exposed individuals involved in the Chernobyl accident met with limited success. Thyroid cancer operations have been successful.

Gastrointestinal Kinetics

The vulnerability of the small intestine to radiation is primarily in the cell renewal system of the intestinal wall. Epithelial cell formation, migration, and loss occur in the crypt and villus structures. Stem cell and proliferating cells move from crypts into the necks of the crypts and bases of the villi. Functionally mature epithelial cells migrate up the villus walls and are extruded at the villus tip. The overall transit time from stem cell to extrusion on the villus for humans is estimated as being 7 to 8 days.

Because of the high turnover rate occurring within the stem cell and proliferating cell compartment of the crypt, marked damage occurs in the region from whole-body radiation doses above the midlethal range. Destruction as well as mitotic inhibition occurs within highly radiosensitive crypt cells within hours after high dosage. Maturing and functional epithelial cells continue to migrate up the villus wall and are extruded, albeit the process is slowed. Shrinkage of villi and morphological changes in mucosal cells occur as new cell production is diminished within the crypts.

Continued loss of epithelial cells in the absence of cell production results in denudation of the intestinal mucosa. Concomitant injury to the microvasculature of the mucosa results in hemorrhage and marked fluid and electrolyte loss contributing to shock. These events normally occur within 1 to 2 weeks after irradiation.

Radiation-Induced Early Transient Incapacitation

Early transient incapacitation (ETI) is associated with very high acute doses of radiation. In humans, with the exception of the bombs dropped at Hiroshima and Nagasaki, it has only occurred during fuel reprocessing, and reactor accidents. The lower limit is probably 20 to 40 Gy. The latent period, a return of partial functionality, is very short, varying from several hours to 1 to 3 days. Subsequently, a deteriorating state of consciousness with vascular instability and death is typical. Convulsions without increased intracranial pressure may or may not occur.

Personnel close enough to a nuclear explosion to develop ETI would die due to blast and thermal effects. However, in nuclear detonations above the atmosphere with essentially no blast, very high fluxes of ionizing radiation may extend out far enough to result in high radiation doses to aircraft crews. Such personnel could conceivably manifest this syndrome, uncomplicated by blast or thermal injury. Also, personnel protected from blast and thermal effects in shielded areas could also sustain such doses. Doses in this range could also result from military operations in a reactor facility or fuel reprocessing plant where personnel are accidentally or deliberately wounded by a nuclear criticality event or close contact with unshielded fission products such as expended fuel.

Time Profile

Acute radiation syndrome (ARS) is a sequence of phased symptoms. Symptoms vary with individual radiation sensitivity, type of radiation, and the radiation dose absorbed. The extent of symptoms will heighten and the duration of each phase will shorten with increasing radiation dose.

Prodromal Phase

The prodrome (or premonitory symptom) is characterized by the relatively rapid onset of nausea, vomiting, and malaise.

This is a nonspecific clinical response to acute radiation exposure. An early onset of symptoms in the absence of associated trauma suggests a large radiation exposure. Radiogenic vomiting may easily be confused with psychogenic vomiting that often results from stress and realistic fear reactions. Use of oral prophylactic antiemetics, such as granisetron (Kytril) and ondansetron (Zofran) may be indicated in situations where high-dose radiological exposure is likely or unavoidable. The purpose of the drug would be to reduce other traumatic injuries after irradiation by mainlining short-term full physical capability.

These medications will diminish the nausea and vomiting in a significant percentage of those personnel exposed and consequently decreases the likelihood of a compromised individual being injured because he was temporarily debilitated. The prophylactic antiemetics do not change the degree of injury due to irradiation and are not radioprotectants. They do diminish the reliability of nausea and emesis as indicators of radiation exposure.

Latent Periods

Following recovery from the prodromal phase, the exposed individual will be relatively symptom-free. The length of this phase varies with the dose. The latent phase is longest preceding the bone-marrow depression of the hematopoietic syndrome and may vary between 2 and 6 weeks.

The latent period is somewhat shorter prior to the gastrointestinal syndrome, lasting from a few days to a week. It is shortest of all preceding the neurovascular syndrome, lasting only a matter of hours. These times are exceedingly variable and may be modified by the presence of other disease or injury. Because of the extreme variability, it is not practical to hospitalize all personnel suspected of having radiation injury early in the latent phase.

Manifest Illness

This phase presents with the clinical symptoms associated with the major organ system injured (marrow, intestinal, neurovascular).

Clinical Acute Radiation Syndrome

Patients who have received doses of radiation between 0.7 and 4 Gy will have depression of bone-marrow function leading to pancytopenia. Changes within the peripheral blood profile will occur as early as 24 hours post-irradiation. Lymphocytes will be depressed most rapidly; other leukocytes and thrombocytes will be depressed somewhat less rapidly.

Decreased resistance to infection and anemia will vary considerably from as early as 10 days to as much as 6 to 8 weeks after exposure. Erythrocytes are least affected due to their useful life span in circulation.

The average time of onset of clinical problems of bleeding and anemia and decreased resistance to infection is 2 to 3 weeks. Even potentially lethal cases of bone-marrow depression may not occur until 6 weeks after exposure. The presence of other injuries will increase the severity and accelerate the time of maximum bone-marrow depression.

The most useful forward laboratory procedure to evaluate marrow depression is the peripheral blood count. A 50 percent drop in lymphocytes within 24 hours indicates significant radiation injury. Bone-marrow studies will rarely be possible under field conditions and will add little information to that which can be obtained from a careful peripheral blood count. Early therapy should prevent nearly all deaths from marrow injury alone.

Higher single gamma-ray doses or radiation (6–8 Gy) will result in the gastrointestinal syndrome, and it will almost always be accompanied by bone-marrow suppression. After a short latent period of a few days to a week or so, the characteristic severe fluid losses, hemorrhage, and diarrhea begin. Derangement of the luminal epithelium and injury to the fine vasculature of the submucosa lead to loss of intestinal mucosa. Peripheral blood counts done on these patients will show the early onset of a severe pancytopenia occurring as a result of the bone-marrow depression. Radiation enteropathy consequently does not result in an inflammatory response.

It must be assumed during the care of all patients that even those with a typical gastrointestinal syndrome may be salvageable. Replacement of fluids and prevention of infection by bacterial transmigration is mandatory.

The neurovascular syndrome is associated only with very high acute doses of radiation (20–40 Gy). Hypotension may be seen at lower doses. The latent period is very short, varying from several hours to 1 to 3 days. The clinical picture is of a steadily deteriorating state of consciousness with eventual coma and death. Convulsions may or may not occur, and there may be little or no indication of increased intracranial pressure. Because of the very high doses of radiation required to cause this syndrome, personnel close enough to a nuclear explosion to receive such high doses would generally be located well within the range of 100 percent lethality due to blast and thermal effects.

A.3   Management Protocol for Acute Radiation Syndrome

The medical management of radiation and combined injuries can be divided into three states: triage, emergency care, and definitive care. During triage, patients are prioritized and rendered immediate lifesaving care. Emergency care includes therapeutics and diagnostics necessary during the first 12 to 24 hours. Definitive care is rendered when final disposition and therapeutic regimens are established.

Effective quality care can be provided both when there are few casualties and a well-equipped facility and when there are many casualties and a worldwide evacuation system. The therapeutic modalities will vary according to current medical knowledge and experience, the number of casualties, available medical facilities, and resources. Recommendations for the treatment of a few casualties may not apply to the treatment of mass casualties because of limited resources. A primary goal should be the evacuation of a radiation casualty prior to the onset of manifest illness.

Prodromal symptoms begin within hours of exposure. They include nausea, vomiting, diarrhea, fatigue, weakness, fever, and headache. The prodromal gastrointestinal symptoms generally do not last longer than 24 to 48 hours after exposure, but a vague weakness and fatigue can persist for an undetermined length of time. The time of onset, severity, and duration of these signs are dose dependent and dose-rate dependent. They can be used in conjunction with white blood cell differential counts to determine the presence and severity of the acute radiation syndrome.

Both the rate and degree of decrease in blood cells are dose dependent. A useful rule of thumb: If lymphocytes have decreased by 50 percent and are less than $1 \times 10^9/l$ (1,000/µl [micro liter]) within 24 to 48 hours, the patient has received at least a moderate dose of radiation. In combined injuries, lymphocytes may be an unreliable indicator. Patients with severe burns and/or trauma to more than one system often develop lymphopenia. These injuries should be assessed by Standard Procedures, keeping in mind that the signs and symptoms of tissue injuries can mimic and obscure those caused by acute radiation effects.

Conventional Therapy for Neutropenia and Infection

The prevention and management of infection is the mainstay of therapy. Antibiotic prophylaxis should only be considered in afebrile patients at the highest risk for infection. These patients have profound neutropenia (<0.1 percent $1 \times 10^9/l$ (100/µl)) that has an expected duration of greater than 7 days. The degree of neutropenia (absolute neutrophils count [ANC] < 100/µl) is the greatest risk factor for developing infection. As the duration of neutropenia increases, the risk of secondary infection such as invasive mycoses also increases. For these reasons, adjuvant therapies such as the use of cytokines will prove invaluable in the treatment of the severely irradiated person.

Prevention of Infection

Initial care of medical casualties with moderate and severe radiation exposure should probably include early institution of measures to reduce pathogen acquisition, with emphasis on low-microbial-content food, acceptable water supplies, frequent hand washing (or wearing of gloves), and air filtration. During the neutropenic period, prophylactic use of selective gut decontamination with antibiotics that suppress aerobes but preserve anaerobes is recommended. The use of sucralfate or prostaglandin analogs may prevent gastric hemorrhage without decreasing gastric activity. When possible, early oral feeding is preferred to intravenous feeding to maintain the immunologic and physiologic integrity of the gut.

Basic Principles

Principle 1: The spectrum of infecting organisms and antimicrobial susceptibility patterns vary both among institutions and over time.

Principle 2: Life-threatening, gram-negative bacterial infections are universal among neutropenic patients, but the prevalence of life-threatening, gram-positive bacterial infections varies greatly.

Principle 3: Current empirical antimicrobial regimens are highly effective for initial management of feverish, neutropenic episodes.

Principle 4: The nidus of infection (that is, the reason the patient is infected) must be identified and eliminated.

Overall Recommendations

A standardized plan for the management of feverish, neutropenic patients must be devised.

Empirical regimens must contain antibiotics broadly active against gram-negative bacteria, but antibiotics directed against the gram-positive bacteria need to be included only in institutions where these infections are prevalent.

No single antimicrobial regimen can be recommended above all others, as pathogens and susceptibility vary with time.

If infection is documented by cultures, the empirical regimen may require adjustment to provide appropriate coverage for the isolate. This should not narrow the antibiotic spectrum.

If the patient defervesces and remains afebrile, the initial regimen should be continued for a minimum of 7 days.

Management of Infection

The management of established or suspected infection (neutropenia and fever) in irradiated persons is similar to that used for other feverish neutropenic patients, such as solid tumor patients receiving chemotherapy. An empirical regimen of antibiotics should be selected, based on the pattern of bacterial susceptibility and nosocomial infections (that is, those arising within) in the particular institution. Broad-spectrum empirical therapy with high doses of one or more antibiotics should be used, avoiding aminoglycosides whenever feasible due to associated toxicities. Therapy should be continued until the patient is afebrile for 24 hours and the ANC is greater than 0.5 x 10^9cells/μl (500 cells /μl) Combination regimens often prove to be more effective than monotherapy. The potential for additivity or synergy should be present in the choice of antibiotics.

For the layman or local emergency manager, all of the above may be overwhelming in technical terms, as it might be to even many medical professionals. Everyone should remember that the U.S. primary source of advice and assistance in treating radiological exposures is maintained by the U.S. Department of Energy (DOE) at Oak Ridge National Laboratory (ORNL) in the form of the staff at REAC/TS.

Appendix B

BIOLOGICAL AGENTS AND THEIR EFFECTS

B.1   BACTERIAL AGENTS

Bacteria generally cause disease in human beings and animals by one of two mechanisms: by invading host tissues and by producing poisons (toxins). Many pathogenic bacteria utilize both mechanisms. The diseases they produce often respond to specific therapy with antibiotics. It is important to distinguish between the disease-causing organism and the name of the disease it causes (in parentheses below). This manual covers several of the bacteria or rickettsiae considered to be potential biological warfare threat agents: *Bacillus anthracis* (anthrax), *Brucella* spp. (brucellosis), *Burkholderia mallei* (glanders), *Burkholderia pseudomallei* (melioidosis), *Yersinia pestis* (plague), *Coxiella burnetii* (Q-fever) and *Francisella tularensis* (tularemia).

B.1.1 ANTHRAX

Signs and Symptoms: Incubation period is generally 1–6 days, although longer periods have been noted. Fever, malaise, fatigue, cough, and mild chest discomfort progresses to severe respiratory distress with dyspnea, diaphoresis, stridor, cyanosis, and shock. Death typically occurs within 24–36 hours after onset of severe symptoms.

Diagnosis: Physical findings are nonspecific. A widened mediastinum may be seen on chest X-rays in later stages of illness. The organism is detectable by Gram stain of the blood and by blood culture late in the course of illness.

Treatment: Although effectiveness may be limited after symptoms are present, high-dose antibiotic treatment with penicillin, ciprofloxacin, or doxycycline should be undertaken. Supportive therapy may be necessary.

Prophylaxis: Oral ciprofloxacin or doxycycline for known or imminent exposure.

An FDA-licensed vaccine is available. Vaccine schedule is 0.5 ml SC at 0, 2, and 4 weeks, then at 6, 12, and 18 months (primary series), followed by annual boosters.

Isolation and Decontamination: Standard Precautions for health care workers. After an invasive procedure or autopsy is performed, the instruments and area used should be thoroughly disinfected with a sporicidal agent (hypochlorite).

Overview

Bacillus anthracis, the causative agent of anthrax, is a gram-positive, sporulating rod. The spores are the usual infective form. Anthrax is primarily a disease of herbivores, with cattle, sheep, goats, and horses being the usual domesticated animal hosts, but other animals may be infected. Humans generally contract the disease when handling contaminated hair, wool, hides, flesh, blood, and excreta of infected animals and from manufactured products such as bone meal. Infection is introduced through scratches or abrasions of the skin, wounds, inhalation of spores, eating insufficiently cooked infected meat, or by biting flies. **The primary concern for intentional infection by this organism is through inhalation after aerosol dissemination of spores. All human populations are susceptible.** The spores are very stable and may remain viable for many years in soil and water. They resist sunlight for varying periods.

History and Significance

Anthrax has been a disease of concern for several millennia, with reports of anthrax-like symptoms recorded in the days of the Pharaohs as well as the ancient Hindus. Anthrax spores were weaponized by the United States in the 1950s and 1960s before the old U.S. offensive program was terminated. Other countries have weaponized this agent or are suspected of doing so. Anthrax bacteria are easy to cultivate and spore production is readily induced. Moreover, the spores are highly resistant to sunlight, heat, and disinfectants—properties that could be advantageous when choosing a bacterial weapon. Iraq admitted to a United Nations inspection team in August of 1991 that it had performed research on the offensive use of *B. anthracis* prior to the Gulf War, and in 1995 Iraq admitted to weaponizing anthrax. A recent defector from the former Soviet Union's biological weapons program revealed that the Soviets had produced anthrax in ton quantities for use as a weapon. This agent could be produced in either a wet or dried form, stabilized for weaponization by an adversary, and delivered as an aerosol cloud either from a line source such as an aircraft flying upwind of friendly positions, or as a point source from a spray device. Coverage of a large ground area could also be theoretically facilitated by multiple spray bomblets disseminated from a missile warhead at a predetermined height above the ground. More contemporaneously, the anthrax "postal attacks" in the United States that left 18 persons infected while killing five are a vivid reminder of the deadly killing power of this disease.

Medical Management

Almost all inhalational anthrax cases in which treatment was begun after patients were significantly symptomatic have been fatal, regardless of treatment. Penicillin has been regarded as the treatment of choice, with 2 million units given intravenously every 2 hours. Tetracyclines and erythromycin have been recommended in penicillin-allergic patients. The vast majority of naturally occurring anthrax strains are sensitive *in vitro* to penicillin. However, penicillin-resistant strains exist naturally, and one has been recovered from a fatal human case.

Moreover, it might not be difficult for an adversary to induce resistance to penicillin, tetracyclines, erythromycin, and many other antibiotics through laboratory manipulation of organisms. All naturally occurring strains tested to date have been sensitive to erythromycin, chloramphenicol, gentamicin, and ciprofloxacin. In the absence of antibiotic sensitivity data, empirical intravenous antibiotic treatment should be instituted at the earliest signs of disease. Military policy currently recommends ciprofloxacin (400 mg IV [intravenously] q 12 hrs) or doxycycline (200 mg IV load, followed by 100 mg IV q 12 hrs) as initial therapy, with penicillin (4 million U IV q 4 hours) as an alternative once sensitivity data is available. Published recommendations from a public health consensus panel recommend ciprofloxacin as initial therapy. Therapy may then be tailored once antibiotic sensitivity is available to penicillin G or doxycycline. Recommended treatment duration is 60 days, and should be changed to oral therapy as clinical condition improves. Supportive therapy for shock, fluid volume deficit, and adequacy of airway may all be needed.

Standard Precautions are recommended for patient care. There is no evidence of direct person-to-person spread of disease from inhalational anthrax. After an invasive procedure or autopsy, the instruments and area used should be thoroughly disinfected with a sporicidal agent. Iodine can be used, but must be used at disinfectant strengths, as antiseptic-strength iodophors are not usually sporicidal. Chlorine, in the form of sodium or calcium hypochlorite, can also be used, but with the caution that the activity of hypochlorites is greatly reduced in the presence of organic material.

Prophylaxis

Vaccine: A licensed vaccine (Anthrax Vaccine Adsorbed) is derived from sterile culture fluid supernatant taken from an attenuated strain. Therefore, the vaccine does not contain live or dead organisms. The vaccination series consists of six 0.5-ml doses SC at 0, 2, and 4 weeks, then 6, 12, and 18 months, followed by yearly boosters. A human efficacy trial in mill workers demonstrated protection against cutaneous anthrax. **There are insufficient data to know its efficacy against inhalational anthrax in humans, although studies in rhesus monkeys indicate that good protection can be afforded after only two doses (15 days apart) for up to 2 years. However, it should be emphasized that the vaccine series should be completed according to the licensed six-dose primary series.** As with all vaccines, the degree of protection depends upon the magnitude of the challenge dose; vaccine-induced protection could presumably be overwhelmed by extremely high spore challenge. Current military policy is to restart the primary vaccine series only if greater than two years elapse between the first and second doses. For all other missed doses, administer the missed dose ASAP and reset the timeline for the series based on the most recent dose.

Contraindications for use of this vaccine include hypersensitivity reaction to a previous dose of vaccine and age younger than 18 or older than 65. Reasons for temporary deferment of the vaccine include pregnancy, active infection with fever, or a course of immune-suppressing drugs such as steroids. Reactogenicity is mild to moderate. Up to 30 percent of recipients may experience mild discomfort at the inoculation site for up to 72 hours (for example, tenderness, erythema, edema, pruritus), fewer experience moderate reactions, while less than 1 percent may experience more severe local reactions, potentially limiting use of the arm for 1–2 days. Modest systemic reactions (for example, myalgia, malaise, low-grade fever) are uncommon, and severe systemic reactions such as anaphylaxis, which precludes additional vaccination, are rare. The vaccine should be stored between 2 and 6°C (refrigerator temperature, not frozen).

Antibiotics: Both military doctrine and a public health consensus panel recommend prophylaxis with ciprofloxacin—Cipro—(500 mg po bid [by mouth twice daily]) as the first-line medication in a situation with anthrax as the presumptive agent. Ciprofloxacin recently became the first medication approved by the FDA for prophylaxis after exposure to a biological weapon (anthrax). Alternatives are doxycycline (100 mg po bid) or amoxicillin (500 mg po q 8 hours), if the strain is susceptible. Should an attack be confirmed as anthrax, antibiotics should be continued for at least 4 weeks, and until all those exposed have received three doses of the vaccine. Those who have already received three doses within 6 months of exposure should continue with their routine vaccine schedule. In the absence of vaccine, chemoprophylaxis should continue for at least 60 days. Upon discontinuation of antibiotics, patients should be closely observed. If clinical signs of anthrax occur, empirical therapy for anthrax is indicated, pending etiologic diagnosis. Optimally, patients should have medical care available upon discontinuation of antibiotics, from a fixed medical care facility with intensive care capabilities and infectious disease consultants.

B.1.2 BRUCELLOSIS

Signs and Symptoms: Illness, when manifest, typically presents with fever, headache, myalgias, arthralgias, back pain, sweats, chills, and generalized malaise. Other manifestations include depression, mental status changes, and osteoarticular findings (that is, sacroileitis, vertebral osteomyelitis). Fatalities are uncommon.

Diagnosis: Diagnosis requires a high index of suspicion, since many infections present as nonspecific febrile illnesses or are asymptomatic. Laboratory diagnosis can be made by blood culture with prolonged incubation. Bone-marrow cultures produce a higher yield. Confirmation requires phage typing, oxidative metabolism, or genotyping procedures. ELISA, followed by Western blot, is available.

Treatment: Antibiotic therapy with doxycycline + rifampin or doxycycline in combination with other medications for 6 weeks is usually sufficient in most cases. More prolonged regimens may be required for patients with complications of meningoencephalitis, endocarditis, or osteomyelitis.

Prophylaxis: There is no human vaccine available against brucellosis, although animal vaccines exist. Chemoprophylaxis is not recommended after possible exposure to endemic disease. Treatment should be considered for high-risk exposure to the veterinary vaccine, inadvertent laboratory exposure, or confirmed biological warfare exposure.

Isolation and Decontamination: Standard Precautions are appropriate for health care workers. Person-to-person transmission has been reported via tissue transplantation and sexual contact. Environmental decontamination can be accomplished with a 0.5 percent hypochlorite solution.

Overview

Brucellosis is one of the world's most important veterinary diseases, and is caused by infection with one of six species of *Brucellae,* a group of gram-negative cocco-baccillary facultative intracellular pathogens. In animals, brucellosis primarily involves the reproductive tract, causing septic abortion and orchitis, which, in turn, can result in sterility. Consequently, brucellosis is a disease of great potential economic impact in the animal husbandry industry. Four species (*B. abortus, B. melitensis, B. suis*, and, rarely, *B. canis*) are pathogenic in humans. Infections in abattoir and laboratory workers suggest that the *Brucellae* are highly infectious via the aerosol route. It is estimated that inhalation of only 10 to 100 bacteria is sufficient to cause disease in humans. Brucellosis has a low mortality rate (5 percent of untreated cases), with rare deaths caused by endocarditis or meningitis. Also, given that the disease has a relatively long and variable incubation period (5–60 days), and that many naturally occurring infections are asymptomatic, its usefulness as a weapon may be diminished. Large aerosol doses, however, may shorten the incubation period and increase the clinical attack rate, and the disease is relatively prolonged, incapacitating, and disabling in its natural form.

History and Significance

Marston described the manifestations of disease caused by *B. melitensis* ("Mediterranean gastric remittent fever") among British soldiers on Malta during the Crimean War. Work by the Mediterranean Fever Commission identified goats as the source, and restrictions on the ingestion of unpasteurized goat milk products soon decreased the number of brucellosis cases among military personnel.

In 1954, *Brucella suis* became the first agent weaponized by the United States at Pine Bluff Arsenal when its offensive biological warfare program was active. *Brucella* weapons, along with the remainder of the U.S. biological arsenal, were destroyed in 1969, when the offensive program was disbanded.

Human brucellosis is now an uncommon disease in the United States, with an annual incidence of 0.5 cases per 100,000 population. Most cases are associated with the ingestion of unpasteurized dairy products, or with abattoir and veterinary work. The disease is, however, highly endemic in Southwest

Asia (annual incidence as high as 128 cases per 100,000 in some areas of Kuwait), thus representing a hazard to military personnel stationed in that theater.

Medical Management

Standard Precautions are adequate in managing brucellosis patients, as the disease is not generally transmissible from person to person. As noted previously, BSL-3 practices should be used when handling suspected *Brucella* cultures in the laboratory because of the danger of inhalation in this setting.

Oral antibiotic therapy alone is sufficient in most cases of brucellosis. Exceptions involve uncommon cases of localized disease, where surgical intervention may be required (for example, valve replacement for endocarditis). A combination of doxycycline 200 mg/d po [by mouth] + rifampin 600 mg/d po is generally recommended. Both drugs should be administered for 6 weeks. Doxycycline 200 mg/d po for 6 weeks in combination with 2 weeks of streptomycin (1 g/d IM [intramuscular]) is an acceptable alternative. Regimens involving doxycycline + gentamicin, TMP/SMX (trimethoprim and sulfamethoxazole) + gentamicin, and ofloxacin + rifampin have also been studied and shown effective. Long-term triple-drug therapy with rifampin, a tetracycline, and an aminoglycoside is recommended by some experts for patients with meningoencephalitis or endocarditis.

Prophylaxis

The risk of endemic brucellosis can be diminished by the avoidance of unpasteurized goat milk and dairy products, especially while traveling in areas where veterinary brucellosis remains common. Live animal vaccines are used widely, and have eliminated brucellosis from most domestic animal herds in the United States, although no licensed human brucellosis vaccine is available.

Chemoprophylaxis is not generally recommended following possible exposure to endemic disease. A 3- to 6-week course of therapy (with one of the regimens discussed) should be considered following a high-risk exposure to veterinary vaccine (such as a needle-stick injury), inadvertent exposure in a laboratory, or exposure in a biological warfare context.

B.1.3 GLANDERS AND MELIOIDOSIS

Signs and Symptoms: Incubation period ranges from 10 to 14 days after inhalation. Onset of symptoms may be abrupt or gradual. Inhalational exposure produces fever (common in excess of 102°F), rigors, sweats, myalgias, headache, pleuritic chest pain, cervical adenopathy, hepatosplenomegaly, and generalized papular/pustular eruptions. Acute pulmonary disease can progress and result in bacteremia and acute septicemic disease. Both diseases are almost always fatal without treatment.

Diagnosis: Methylene blue or Wright stain of exudates may reveal scant small bacilli with a safety-pin bipolar appearance. Standard cultures can be used to identify both *B. mallei* and *B. pseudomallei*. Chest X-rays may show miliary lesions, small multiple lung abscesses, or infiltrates involving upper lungs, with consolidation and cavitation. Leukocyte counts may be normal or elevated. Serologic tests can help confirm diagnosis, but low titers or negative serology does not exclude the diagnosis.

Treatment: Therapy will vary with the type and severity of the clinical presentation. Patients with localized disease may be managed with oral antibiotics for a duration of 60–150 days.

More severe illness may require parenteral therapy and more prolonged treatment.

Prophylaxis: Currently, no pre-exposure or post-exposure prophylaxis is available.

Isolation and Decontamination: Standard Precautions for health care workers. Person-to-person airborne transmission is unlikely, although secondary cases may occur through improper handling of infected secretions. Contact Precautions are indicated while caring for patients with skin involvement. Environmental decontamination using a 0.5 percent hypochlorite solution is effective.

Overview

The causative agents of glanders and melioidosis are *Burkholderia mallei* and *Burkholderia pseudomallei,* respectively. Both are gram-negative bacilli with a "safety-pin" appearance on microscopic examination. Both pathogens affect domestic and wild animals, which, like humans, acquire the diseases from inhalation or contaminated injuries.

Burkholderia mallei is primarily noted for producing disease in horses, mules, and donkeys. In the past humans have seldom been infected, despite frequent and often close contact with infected animals. This may be the result of exposure to low concentrations of organisms from infected sites in ill animals and because strains virulent for equids are often less virulent for man. There are four basic forms of disease in horses and man. The acute forms are more common in mules and donkeys, with death typically occurring 3 to 4 weeks after illness onset. The chronic form of the disease is more common in horses and causes generalized lymphadenopathy, multiple skin nodules that ulcerate and drain, and induration, enlargement, and nodularity of regional lymphatics on the extremities and in other areas. The lymphatic thickening and induration has been called farcy. Human cases have occurred primarily in veterinarians, horse and donkey caretakers, and abattoir workers.

Burkholderia pseudomallei is widely distributed in many tropical and subtropical regions. The disease is endemic in Southeast Asia and northern Australia. In northeastern Thailand, B. pseudomallei is one of the most common causative agents of community-acquired septicemia. Melioidosis presents in humans in several distinct forms, ranging from a subclinical illness to an overwhelming septicemia, with a 90 percent mortality rate and death within 24–48 hours after onset. Also, melioidosis can reactivate years after primary infection and result in chronic and life-threatening disease.

These organisms spread to humans by invading the nasal, oral, and conjunctival mucous membranes, by inhalation into the lungs, and by invading abraded or lacerated skin. Aerosols from cultures have been observed to be highly infectious to laboratory workers. BSL-3 containment practices are required when working with these organisms in the laboratory. Since aerosol spread is efficient, and there is no available vaccine or reliable therapy, B. mallei and B. pseudomallei have both been viewed as potential biological warfare agents.

History and Significance

Despite the efficiency of spread in a laboratory setting, glanders has only been a sporadic disease in man, and no epidemics of human disease have been reported. There have been no naturally acquired cases of human glanders in the United States in over 61 years. Sporadic cases continue to occur in Asia, Africa, the Middle East, and South America. During World War I, glanders was believed to have been spread deliberately by agents of the Central Powers to infect large numbers of Russian horses and mules on the Eastern Front. This had an effect on troop and supply convoys as well as on artillery movement, which were dependent on horses and mules. Human cases in Russia increased with the infections during and after World War I. The Japanese deliberately infected horses, civilians, and prisoners of war with *B. mallei* at the Pinfang (China) Institute during World War II. The United States studied this agent as a possible biological warfare weapon in 1943–1944 but did not weaponize it. The former Soviet Union is believed to have been interested in *B. mallei* as a potential biological warfare agent after World War II. The low transmission rates of *B. mallei* to man from infected horses is exemplified by the fact that in China, during World War II, 30 percent of tested horses were positive for glanders, but human cases were rare. In Mongolia, 5–25 percent of tested animals were reactive to *B. mallei*, but no human cases were seen. *Burkholderia mallei* exists in nature only in infected susceptible hosts and is not found in water, soil, or plants.

In contrast, melioidosis is widely distributed in the soil and water in the tropics, and remains endemic in certain parts of the world, even to this day. It is one of the few genuinely tropical diseases that are well established in Southeast Asia and northern Australia. As a result of its long incubation period, it could be unknowingly imported.

Burkholderia pseudomallei was also studied by the United States as a potential biological warfare agent, but was never weaponized. It has been reported that the former Soviet Union was experimenting with *B. pseudomallei* as a biological warfare agent.

Medical Management
Standard Precautions should be used to prevent person-to-person transmission in proven or suspected cases. The recom-

mended therapy will vary with the type and severity of the clinical presentation. **The following oral regimens have been suggested for localized disease: amoxicillin/clavulanate 60 mg/kg/day in three divided doses; tetracycline 40 mg/kg/day in three divided doses; or trimethoprim/sulfa (TMP 4 mg/kg/day, sulfa 20 mg/kg/day) in two divided doses. The duration of treatment should be for 60–150 days.**

If the patient has localized disease with signs of mild toxicity, then a combination of two of the oral regimens is recommended for a duration of 30 days, followed by monotherapy with either amoxicillin/clavulanate or TMP/sulfa for 60–150 days. If extrapulmonary suppurative disease is present, then therapy should continue for 6–12 months. Surgical drainage of abscesses may be required.

For severe disease, parenteral therapy with ceftazidime 120 mg/kg/day in three divided doses combined with TMP/sulfa (TMP 8 mg/kg/day, sulfa 40 mg/kg/day) in four divided doses for 2 weeks is recommended, followed by oral therapy for 6 months.

Other antibiotics that have been effective in experimental infection in hamsters include doxycycline, rifampin, and ciprofloxacin. The limited number of infections in humans has precluded therapeutic evaluation of most of the antibiotic agents; therefore, most antibiotic sensitivities are based on animal and *in vitro* studies. Various isolates have markedly different antibiotic sensitivities; therefore, each isolate should be tested for its own resistance pattern.

Prophylaxis

Vaccine: There is no vaccine available for human use.

Antibiotics: Post-exposure chemoprophylaxis may be tried with TMP/SMX (the antibiotics trimethoprim and sulfamethoxazole).

B.1.4 PLAGUE

Signs and Symptoms: Pneumonic plague begins after an incubation period of 1–6 days, with high fever, chills, headache, and malaise, followed by cough (often with hemoptysis), progressing rapidly to dyspnea, stridor, cyanosis, and death. Gastrointestinal symptoms are often present. Death results from respiratory failure, circulatory collapse, and a bleeding diathesis. Bubonic plague, featuring high fever, malaise, and painful lymph nodes (buboes), may progress spontaneously to the septicemic form (septic shock, thrombosis) or to the pneumonic form.

Diagnosis: Suspect plague if large numbers of previously healthy individuals develop fulminant gram-negative pneumonia, especially if hemoptysis is present. Presumptive diagnosis can be made by Gram, Wright, Giemsa, or Wayson stain of blood, sputum, cerebrospinal fluid, or lymph node aspirates. Definitive diagnosis requires culture of the organism from those sites. Immunodiagnosis is also helpful.

Treatment: Early administration of antibiotics is critical, as pneumonic plague is invariably fatal if antibiotic therapy is delayed more than 1 day after the onset of symptoms. Choose one of the following: streptomycin, gentamicin, ciprofloxacin, or doxycycline for 10–14 days. Chloramphenicol is the drug of choice for plague meningitis.

Prophylaxis: For asymptomatic persons exposed to a plague aerosol or to a patient with suspected pneumonic plague, give doxycycline 100 mg orally twice daily for 7 days or the duration of risk of exposure plus one week. Alternative antibiotics include ciprofloxacin, tetracycline, or chloramphenicol. No vaccine is currently available for plague prophylaxis. The previously available licensed, killed vaccine was effective against bubonic plague, but not against aerosol exposure.

Isolation and Decontamination: Use Standard Precautions for bubonic plague, and Respiratory Droplet Precautions for suspected pneumonic plague. *Yersinia pestis* can survive in the environment for varying periods, but is susceptible to heat, disinfectants, and exposure to sunlight. Soap and water is effective if decontamination is needed. Take measures to prevent local disease cycles if vectors (fleas) and reservoirs (rodents) are present.

Overview

Yersinia pestis is a rod-shaped, nonmotile, nonsporulating, gram-negative bacterium of the family *Enterobacteraceae*. It causes plague, a disease of rodents (for example, rats, mice, ground squirrels). Fleas that live on the rodents can transmit the bacteria to humans, who then suffer from the bubonic form of plague. The bubonic form may progress to the septicemic and/or pneumonic forms. Pneumonic plague would be the predominant form after a purposeful aerosol dissemination. All human populations are susceptible. Recovery from the disease is followed by temporary immunity. The organism remains viable in water, moist soil, and grains for several weeks. At near freezing temperatures, it will remain alive from months to years but is killed by 15 minutes of exposure to 55°C. It also remains viable for some time in dry sputum, flea feces, and buried bodies but is killed within several hours of exposure to sunlight.

History and Significance

The United States worked with *Y. pestis* as a potential bio-warfare agent in the 1950s and 1960s before the old offensive bio-warfare program was terminated, but other countries are suspected of weaponizing this organism. The former Soviet Union had more than 10 institutes and thousands of scientists who worked with plague. During World War II, Unit 731, of the Japanese Army, reportedly released plague-infected fleas from aircraft over Chinese cities. This method was cumbersome and unpredictable. The U.S. and Soviet Union developed the more reliable and effective method of aerosolizing the organism. **The interest in the terrorist potential of plague was brought to light in 1995 when Larry Wayne Harris was arrested in Ohio for the illicit procurement of a _Y. pestis_ culture through the mail. The contagious nature of pneumonic plague makes it particularly dangerous as a biological weapon.**

Medical Management
Use Standard Precautions for bubonic plague patients. Suspected pneumonic plague cases require strict isolation with Droplet Precautions for at least 48 hours of antibiotic therapy, or until sputum cultures are negative in confirmed cases. If competent vectors (fleas) and reservoirs (rodents) are present, measures must be taken to prevent local disease cycles. These might include, but are not limited to, use of flea insecticides, rodent control measures (after or during flea control), and flea barriers for patient care areas.

Streptomycin, gentamicin, doxycycline, and chloramphenicol are highly effective, if begun early. Plague pneumonia is almost always fatal if treatment is not initiated within 24 hours of the onset of symptoms. Dosage regimens are as follows: streptomycin, 30 mg/kg/day IM in two divided doses; gentamicin, 5mg/kg IV once daily, or 2 mg/kg loading dose followed by 1.75 mg/kg IV every 8 hours; doxycycline, 200 mg initially, followed by 100 mg every 12 hours. Duration of therapy is 10 to 14 days. While the patient is typically afebrile after 3 days, the extra week of therapy prevents relapses. Results obtained from laboratory animal, but not human, experience indicate that quinolone antibiotics, such as ciprofloxacin and ofloxacin, may also be effective. Recommended dosage of ciprofloxacin is 400mg IV twice daily. Chloramphenicol, 25 mg/kg IV loading dose followed by 15 mg/kg IV four times daily x 10–14 days, is required for the treatment of plague meningitis.

Usual supportive therapy includes IV crystalloids and hemodynamic monitoring. Although low-grade DIC (disseminated intravascular coagulation) may occur, clinically significant hemorrhage is uncommon, as is the need to treat with heparin. Endotoxic shock is common, but pressor agents are rarely needed. Finally, buboes rarely require any form of local care, but instead recede with systemic antibiotic therapy. In fact, incision and drainage poses a risk to others in contact with the patient; aspiration is recommended for diagnostic purposes and may provide symptomatic relief.

Prophylaxis

Vaccine: No vaccine is currently available for prophylaxis of plague. A licensed, killed whole-cell vaccine was available in the United States from 1946 until November 1998. It offered protection against bubonic plague, but was not effective against aerosolized Y. *pestis*. Presently, an F1-V antigen (fusion protein) vaccine is in development at USAMRIID. It protected mice for a year against an inhalational challenge, and is now being tested in primates.

Antibiotics: Face-to-face contacts (within 2 meters) of patients with pneumonic plague or persons possibly exposed to a plague aerosol in a plague biological warfare attack) should be given antibiotic prophylaxis for 7 days or the duration of risk of exposure plus 7 days. If fever or cough occurs in these individuals, treatment with antibiotics should be started. **Because of oral administration and relative lack of toxicity, the choice of antibiotic for prophylaxis is doxycycline 100 mg orally**

twice daily. Ciprofloxacin 500 mg orally twice daily has also shown to be effective in preventing disease in exposed mice, and may be more available in a wartime setting as it is also distributed in blister-packs for anthrax post-exposure prophylaxis. Tetracycline, 500mg orally four times daily, and chloramphenicol, 25 mg/kg orally four times daily, are acceptable alternatives. Contacts of bubonic plague patients need only be observed for symptoms for a week. If symptoms occur, start treatment antibiotics.

B.1.5 Q-FEVER

Signs and Symptoms: Fever, cough, and pleuritic chest pain may occur as early as 10 days after exposure. Patients are not generally critically ill, and the illness lasts from 2 days to 2 weeks.

Diagnosis: Q-fever is not a clinically distinct illness and may resemble a viral illness or other types of atypical pneumonia. The diagnosis is confirmed serologically.

Treatment: Q-fever is generally a self-limited illness even without treatment, but tetracycline or doxycycline should be given orally for 5 to 7 days to prevent complications of the disease. Q-fever endocarditis (rare) is much more difficult to treat.

Prophylaxis: Chemoprophylaxis begun too early during the incubation period may delay but not prevent the onset of symptoms. Therefore, tetracycline or doxycycline should be started 8–12 days post exposure and continued for 5 days. This regimen has been shown to prevent clinical disease. An inactivated whole-cell investigational (IND) vaccine is effective in eliciting protection against exposure, but severe local reactions to this vaccine may be seen in those who already possess immunity. Therefore, an intradermal skin test is recommended to detect presensitized or immune individuals.

Isolation and Decontamination: Standard Precautions are recommended for health care workers. Person-to-person transmission is rare. Patients exposed to Q-fever by aerosol do not present a risk for secondary contamination or re-aerosolization of the organism. Decontamination is accomplished with soap and water or a 0.5 percent chlorine solution on the infected person. The M291 skin decontamination kit will *not* neutralize the organism.

Overview

The endemic form of Q-fever is a zoonotic disease caused by the rickettsia, *Coxiella burnetii*. Its natural reservoirs are sheep, cattle, goats, dogs, cats, and birds. The organism grows to especially high concentrations in placental tissues. The infected animals do not develop the disease, but do shed large numbers of the organisms in placental tissues and body fluids, including milk, urine, and feces. Exposure to infected animals at parturition is an important risk factor for endemic disease. Humans acquire the disease by inhalation of aerosols contaminated with the organisms. Farmers and abattoir workers are at greatest risk occupationally. A biological warfare attack with Q-fever would cause a disease similar to that occurring naturally. Q-fever is also a significant hazard in laboratory personnel who are working with the organism.

History and Significance

Q-fever was first described in Australia and called "Query fever" because the causative agent was initially unknown. *Coxiella burnetii*, discovered in 1937, is a rickettsial organism that is resistant to heat and desiccation and highly infectious by the aerosol route. A single inhaled organism may produce clinical illness. For all of these reasons, Q-fever could be used by an adversary as an incapacitating biological warfare agent.

Medical Management

Standard Precautions are recommended for health care workers. Most cases of acute Q-fever will eventually resolve without antibiotic treatment, but all suspected cases of Q-fever should be treated to reduce the risk of complications. **Tetracycline 500 mg every 6 hours or doxycycline 100 mg every 12 hours for 5–7 days will shorten the duration of illness, and fever usually disappears within one to two days after treatment is begun.** Ciprofloxacin and other quinolones are active *in vitro* and should be considered in patients unable to take tetracycline or doxycycline. Successful treatment of Q-fever endocarditis is much more difficult. Tetracycline or doxycycline given in combination with trimethoprim-sulfamethoxazole (TMP-SMX) or rifampin for 12 months or longer has been successful in some cases. However, valve replacement is often required to achieve a cure.

Prophylaxis

Vaccine: A formalin-inactivated whole-cell IND vaccine is available for immunization of at-risk personnel on an investigational basis, although a Q-fever vaccine is licensed in Australia. Vaccination with a single dose of this killed suspension of C. *burnetii* provides complete protection against naturally occurring Q-fever, and greater than 95 percent protection against aerosol exposure. Protection lasts for at least 5 years. Administration of this vaccine in immune individuals may cause severe local induration, sterile abscess formation, and even necrosis at the inoculation site. This observation led to the development of an intradermal skin test using 0.02 mg of specific formalin-killed whole-cell vaccine to detect presensitized or immune individuals.

Antibiotics: Chemoprophylaxis using tetracycline 500 mg every 6 hours or doxycycline 100 mg every 12 hours for 5–7 days is effective if begun 8–12 days post exposure. Chemoprophylaxis is not effective and may only prolong the onset of disease if given immediately (1 to 7 days) after exposure.

B.1.6 TULAREMIA

Signs and Symptoms: Ulceroglandular tularemia presents with a local ulcer and regional lymphadenopathy, fever, chills, headache, and malaise. Typhoidal tularemia presents with fever, headache, malaise, substernal discomfort, prostration, weight loss, and a nonproductive cough.

Diagnosis: For clinical diagnosis, physical findings are usually nonspecific. Chest X-ray may reveal a pneumonic process, mediastinal lymphadenopathy, or pleural effusion. Routine culture is possible but difficult. The diagnosis can be established retrospectively by serology.

Treatment: Administration of antibiotics (streptomycin or gentamicin) with early treatment is very effective.

Prophylaxis: A live, attenuated vaccine is available as an investigational new drug. It is administered once by scarification. A 2-week course of tetracycline is effective as prophylaxis when given after exposure.

Isolation and Decontamination: Standard Precautions for health care workers. Organisms are relatively easy to render harmless by mild heat (55°C for 10 minutes) and standard disinfectants.

Overview

Francisella tularensis, the causative agent of tularemia, is a small, aerobic nonmotile, gram-negative coccobacillus. Tularemia (also known as rabbit fever and deerfly fever) is a disease that humans typically acquire after skin or mucous membrane contact with tissues or body fluids of infected animals, or from bites of infected ticks, deerflies, or mosquitoes. Less commonly, inhalation of contaminated dusts or ingestion of contaminated foods or water may produce clinical disease. Respiratory exposure by aerosol would typically cause typhoidal or pneumonic tularemia. *Francisella tularensis* can remain viable for weeks in water, soil, carcasses, and hides, and for years in frozen rabbit meat. It is resistant for months to temperatures of freezing and below. It is easily killed by heat and disinfectants.

History and Significance

Tularemia was recognized in Japan in the early 1800s and in Russia in 1926. In the early 1900s, American workers investigating suspected plague epidemics in San Francisco isolated the organism and named it *Bacterium tularense* after Tulare County, California, where the work was performed. Dr. Edward Francis, of the U.S. Public Health Service, established the cause of deerfly fever as *Bacterium tularense* and subsequently devoted his life to researching the organism and disease, hence the organism was later renamed *Francisella tularensis.*

Francisella tularensis was weaponized by the United States in the 1950s and 1960s during the U.S. offensive bio-warfare program, and other countries are suspected to have weaponized this agent. This organism could potentially be stabilized for weaponization by an adversary and theoretically produced in either a wet or dried form, for delivery against

U.S. forces in a similar fashion to the other bacteria discussed in this manual.

Medical Management

Since there is no known human-to-human transmission, neither isolation nor quarantine is required, since Standard Precautions are appropriate for care of patients with draining lesions or pneumonia. Strict adherence to the drainage/secretion recommendations of Standard Precautions is required, especially for draining lesions, and for the disinfection of soiled clothing, bedding, equipment, etc. Heat and disinfectants easily inactivate the organism.

Appropriate therapy includes one of the following antibiotics:

* Gentamicin 3–5 mg/kg IV daily for 10 to 14 days
* Ciprofloxacin 400 mg IV every 12 hours, switch to oral ciprofloxacin (500 mg every 12 hours) after the patient is clinically improved; continue for completion of a 10- to 14-day course of therapy
* Ciprofloxacin 750 mg orally every 12 hours for 10 to 14 days
* Streptomycin 7.5–10 mg/kg IM every 12 hours for 10 to 14 days

Streptomycin has historically been the drug of choice for tularemia; however, since it may not be readily available immediately after a large-scale biological warfare attack, gentamicin and other alternative drugs should be considered first. Requests for streptomycin should be directed to the Roerig Streptomycin Program at Pfizer Pharmaceuticals in New York (800-254-4445). Another concern is that a fully virulent streptomycin-resistant strain of *F. tularensis* was developed during the 1950s and it is presumed that other countries have obtained it. The strain was sensitive to gentamicin. Gentamicin offers the advantage of providing broader coverage for gram-negative bacteria and may be useful when the diagnosis of tularemia is considered but in doubt.

In a recent study of treatment in 12 children with ulceroglandular tularemia, ciprofloxacin was satisfactory for outpatient treatment (*Pediatric Infectious Disease Journal,* 2000; 19:449–453). Tetracycline and chloramphenicol are also effective antibiotics; however, they are associated with significant relapse rates.

Prophylaxis

Vaccine: An investigational live-attenuated vaccine (Live Vaccine Strain, LVS), which is administered by scarification, has been given to more than 5,000 persons without significant adverse reactions and prevents typhoidal and ameliorates ulceroglandular forms of laboratory-acquired tularemia. Aerosol challenge tests in laboratory animals and human volunteers have demonstrated significant protection. As with all vaccines, the degree of protection depends upon the magnitude of the challenge dose. Vaccine-induced protection could be overwhelmed by extremely high doses of the tularemia bacteria.

Immunoprophylaxis: There is no passive immunoprophylaxis (that is, immune globulin) available for pre- or post-exposure management of tularemia.

Pre-exposure prophylaxis: Chemoprophylaxis given for anthrax or plague (ciprofloxacin, doxycycline) may confer protection against tularemia, based on *in vitro* susceptibilities.

Post-exposure prophylaxis: A 2-week course of antibiotics is effective as post-exposure prophylaxis when given within 24 hours of aerosol exposure from a biological warfare attack, using one of the following regimens:

Ciprofloxacin 500 mg orally every 12 hours for 2 weeks
Doxycycline 100 mg orally every 12 hours for 2 weeks
Tetracycline 500 mg orally every 6 hours for 2 weeks

Chemoprophylaxis is not recommended following potential natural exposures (tick bite, rabbit, or other animal exposures).

B.2   VIRAL AGENTS

Viruses are the simplest microorganisms and consist of a nucleocapsid protein coat containing genetic material, either RNA or DNA. In some cases, the viral particle is also surrounded by an outer lipid layer. Viruses are much smaller than bacteria. Viruses are intracellular parasites and lack a system for their own metabolism; therefore, they are dependent on the synthetic machinery of their host cells. This means that viruses, unlike the bacteria, cannot be cultivated in synthetic nutritive solutions, but require living cells in order to multiply. The host cells can be from humans, animals, plants, or bacteria. Every virus requires its own special type of host cell for multiplication, because a complicated interaction occurs between the cell and virus. Virus-specific host cells can be cultivated in synthetic nutrient solutions and then infected with the virus in question. Another common way of cultivating viruses is to grow them on chorioallantoic membranes (from fertilized eggs). The cultivation of viruses is expensive, demanding, and time-consuming. A virus typically brings about changes in the host cell that eventually lead to cell death. This manual covers three types of viruses, which could potentially be employed as biological warfare agents: smallpox, alphaviruses (for example, Venezuelan equine encephalitis), and hemorrhagic fever viruses.

B.2.1 SMALLPOX

Signs and Symptoms: Clinical manifestations begin acutely with malaise, fever, rigors, vomiting, headache, and backache. Two to three days later lesions appear, which quickly progress from macules to papules, and eventually to pustular vesicles. They are more abundant on the extremities and face, and develop synchronously.

Diagnosis: Neither electron nor light microscopy is capable of discriminating variola from vaccinia, monkeypox, or cowpox. The new PCR diagnostic techniques may be more accurate in discriminating between variola and other orthopoxviruses.

Treatment: At present there is no effective chemotherapy, and treatment of a clinical case remains supportive.

Prophylaxis: Immediate vaccination or revaccination should be undertaken for all individuals exposed.

Isolation and Decontamination: Droplet and Airborne Precautions should be followed for a minimum of 17 days following exposure for all contacts. Patients should be considered infectious until all scabs separate and must be quarantined during this period. In the civilian setting, strict quarantine of asymptomatic contacts may prove to be impractical and impossible to enforce. A reasonable alternative would be to require contacts to check their temperatures daily. Any fever above 38°C (101°F) during the 17-day period following exposure to a confirmed case would suggest the development of smallpox. The contact should then be isolated immediately, preferably at home, until smallpox is either confirmed or ruled out, and remain in isolation until all scabs separate.

Overview

Smallpox is caused by the *Orthopox* virus, variola, which occurs in at least two strains, variola major and the milder dis-

ease, variola minor. **Despite the global eradication of smallpox and continued availability of a vaccine, the potential weaponization of variola continues to pose a military threat.** This threat can be attributed to the aerosol infectivity of the virus, the relative ease of large-scale production, and an increasingly *Orthopoxvirus*-naive populace. Although the fully developed cutaneous eruption of smallpox is unique, earlier stages of the rash could be mistaken for varicella. Secondary spread of infection constitutes a nosocomial hazard from the time of onset of a smallpox patient's exanthem until scabs have separated. Quarantine with respiratory isolation should be applied to secondary contacts for 17 days post-exposure. Vaccinia vaccination and vaccinia immune globulin each possess some efficacy in post-exposure prophylaxis.

History and Significance

Endemic smallpox was declared eradicated in 1980 by the World Health Organization (WHO). Although two WHO-approved repositories of variola virus remain at the Centers for Disease Control and Prevention (CDC) in Atlanta and the Institute for Viral Preparations in Moscow, the extent of clandestine stockpiles in other parts of the world remains unknown. The United States stopped vaccinating its military population in 1989 and civilians in the early 1980s. These populations are now susceptible to variola major, although recruits immunized in 1989 may retain some degree of immunity. In the eighteenth century, variola may have been used by the British Army against Native Americans by giving them contaminated blankets from the beds of smallpox victims. Japan considered the use of smallpox as a biological warfare weapon in World War II and it has been considered as a possible threat agent against U.S. forces for many years. In addition, the former Soviet Union is reported to have produced and stockpiled massive quantities of the virus for use as a biological weapon. It is not known whether these stockpiles still exist in Russia.

Medical Management

Medical personnel must be prepared to recognize a vesicular exanthem in possible bio-warfare theaters as potentially variola, and to initiate appropriate countermeasures. Any confirmed case of smallpox should be considered an international emergency with immediate report made to public health authorities. Droplet and Airborne Precautions are needed for a minimum of 17 days following exposure for *all* persons in direct contact with the index case, especially the unvaccinated. In the civilian setting, strict quarantine of asymptomatic contacts may prove to be impractical and impossible to enforce. A reasonable alternative would be to require contacts to check their temperatures daily. Any fever above 38°C (101°F) during the 17-day period following exposure to a confirmed case would suggest the development of smallpox. The contact should then be isolated immediately, preferably at home, until smallpox is either confirmed or ruled out, and remain in isolation until all scabs separate. Patients should be considered infectious until all scabs separate. Immediate vaccination or revaccination should also be undertaken for all personnel exposed to either weaponized variola virus or a clinical case of smallpox.

The potential for airborne spread to other than close contacts is controversial. In general, close person-to-person contact is required for transmission to reliably occur. Nevertheless, variola's potential in low relative humidity for airborne dissemination was alarming in two hospital outbreaks. Smallpox patients were infectious from the time of onset of their eruptive exanthem, most commonly from days 3–6 after onset of fever. Infectivity was markedly enhanced if the patient manifested a cough. Indirect transmission via contaminated bedding or other fomites was infrequent. Some close contacts harbored virus in their throats without developing disease, and hence might have served as a means of secondary transmission.

Vaccination with a verified clinical "take" (vesicle with scar formation) within the past 3 years is considered to render a person immune to smallpox. However, given the difficulties and uncertainties under wartime conditions of verifying the adequacy of troops' prior vaccination, routine revaccination of all potentially exposed personnel would seem prudent if there existed a significant prospect of smallpox exposure.

Antivirals for use against smallpox are under investigation. Cidofovir has been shown to have significant *in vitro* and *in vivo* activity in experimental animals. Whether it would offer benefit superior to immediate post-exposure vaccination in humans has not been determined.

Prophylaxis

Vaccine: Smallpox vaccine (vaccinia virus) is most often administered by intradermal inoculation with a bifurcated needle, a process that became known as scarification because of the permanent scar that resulted. Vaccination after exposure to weaponized smallpox or a case of smallpox may prevent or ameliorate disease if given as soon as possible and preferably within 7 days after exposure. A vesicle typically appears at the vaccination site 5–7 days post-inoculation, with surrounding erythema and induration. The lesion forms a scab and gradually heals over the next 1–2 weeks.

Side effects include low-grade fever and axillary lymphadenopathy. The attendant erythema and induration of the vaccination vesicle are frequently misdiagnosed as bacterial superinfection. More severe first-time vaccine reactions include secondary inoculation of the virus to other sites such as the face, eyelid, or other persons (~6 of 10,000 vaccinations), and generalized vaccinia, which is a systemic spread of the virus to produce mucocutaneous lesions away from the primary vaccination site (~3 in 10,000 vaccinations).

Vaccination is *contraindicated* in the following conditions: immunosuppression, HIV infection, history or evidence of eczema, or current household, sexual, or other close physical contact with person(s) possessing one of these conditions. In addition, vaccination should not be performed during pregnancy. Despite these caveats, most authorities state that, with the exception of significant impairment of systemic immunity, there are no absolute contraindications to *post-exposure* vaccination of a person who experiences *bona fide* exposure to variola. However, concomitant VIG administration is recommended for pregnant and eczematous persons in such circumstances.

Passive Immunoprophylaxis: Vaccinia Immune Globulin (VIG) is generally indicated for treatment of complications to the smallpox (vaccinia) vaccine, and should be available when administering vaccine. Limited data suggest that vaccinia immune globulin may be of value in post-exposure prophylaxis of smallpox when given within the first week following exposure, and concurrently with vaccination. Vaccination alone is recommended for those without contraindications to the vaccine. If greater than one week has elapsed after exposure, administration of both products, if available, is reasonable. The dose for prophylaxis or treatment is 0.6 ml/kg intramuscularly. Due to the large volume (42 ml in a 70-kg person), the dose would be given in multiple sites over 24–36 hours.

B.2.2 VENEZUELAN EQUINE ENCEPHALITIS

Signs and Symptoms: Incubation period 1–6 days. Acute systemic febrile illness with encephalitis develops in a small percentage (4 percent of children; less than 1 percent of adults). Generalized malaise, spiking fevers, rigors, severe headache, photophobia, and myalgias for 24–72 hours result. Nausea, vomiting, cough, sore throat, and diarrhea may follow. Full recovery from malaise and fatigue takes 1–2 weeks. The incidence of CNS disease and associated morbidity and mortality would be much higher after a biological warfare attack.

Diagnosis: Clinical and epidemiological diagnosis. Physical findings are nonspecific. The white blood cell count may show a striking leukopenia and lymphopenia. Virus isolation may be made from serum, and in some cases throat swab specimens. Both neutralizing and IgG antibody in paired sera or Venezuelan equine encephalitis (VEE) specific IgM present in a single serum sample indicate recent infection.

Therapy: Treatment is supportive only. Treat uncomplicated VEE infections with analgesics to relieve headache and myalgia. Patients who develop encephalitis may require anticonvulsants and intensive supportive care to maintain fluid and electrolyte balance, ensure adequate ventilation, and avoid complicating secondary bacterial infections.

Prophylaxis: A live, attenuated vaccine is available as an investigational new drug. A second, formalin-inactivated, killed vaccine is available for boosting antibody titers in those initially receiving the first vaccine. No post-exposure immuno-prophylaxis. In experimental animals, alpha-interferon and the interferon-inducer poly-ICLC have proven highly effective as post-exposure prophylaxis. There are no human clinical data.

Isolation and Decontamination: Patient isolation and quarantine are not required. Standard Precautions augmented with vector control while the patient is febrile. There is no evidence of direct human-to-human or horse-to-human transmission. The virus can be destroyed by heat (80°C for 30 minutes) and standard disinfectants.

Overview

The Venezuelan equine encephalitis (VEE) virus complex is a group of eight mosquito-borne alphaviruses that are endemic in northern South America and Trinidad and cause rare cases of human encephalitis in Central America, Mexico, and Florida. These viruses can cause severe diseases in humans and Equidae (horses, mules, burros, and donkeys). Natural infections are acquired by the bites of a wide variety of mosquitoes. Equidae serve as amplifying hosts and source of mosquito infection.

Western and eastern equine encephalitis viruses are similar to the VEE complex, are often difficult to distinguish clinically, and share similar aspects of transmission and epidemiology. The human infective dose for VEE is considered to be 10–100 organisms, which is one of the principal reasons that VEE is considered a militarily effective biological warfare agent. Neither the population density of infected mosquitoes nor the aerosol concentration of virus particles has to be great to allow significant transmission of VEE in such an attack. There is no evidence of direct human-to-human or horse-to-human transmission. Natural aerosol transmission is not known to occur. VEE particles are not considered stable in the environment, and are thus not as persistent as the bacteria responsible for Q-fever, tularemia, or anthrax. Heat and standard disinfectants can easily kill the VEE virus complex.

History and Significance

Between 1969 and 1971, an epizootic of a "highly pathogenic strain" of VEE emerged in Guatemala, moved through Mexico, and entered Texas in June 1971. This strain was virulent in both equine species and humans. In Mexico, there were 8,000–10,000 equine deaths, "tens of thousands" of equine cases, and 17,000 human cases (no human deaths). Over 10,000 horses in Texas died. Once the Texas border was breached, a national emergency was declared and resources were mobilized to vaccinate equines in 20 states (95 percent of all horses and donkeys were vaccinated, over 3.2 million animals), establish equine quarantines, and control mosquito populations with broad-scale insecticide use in the Rio Grande Valley and along the Gulf Coast. A second VEE outbreak in 1995 in Venezuela and Colombia involved over 75,000 human cases and over 20 deaths.

VEE is better characterized than the eastern (EEE) or western (WEE) forms, primarily because it was tested as a biological warfare agent during the U.S. offensive program in the 1950s and 1960s. Other countries have also been or are suspected to have weaponized this agent. In compliance with President Nixon's National Security Decision No. 35 of November 1969 to destroy the biological warfare microbial stockpile, all existing stocks of VEE in the United States were publicly destroyed.

These viruses could theoretically be produced in large amounts in either a wet or dried form by relatively unsophisticated and inexpensive systems. This form of the VEE virus complex could be intentionally disseminated as an aerosol and would be highly infectious. It could also be spread by the purposeful dissemination of infected mosquitoes, which can probably transmit the virus throughout their lives. The VEE complex is relatively stable during the storage and manipulation procedures necessary for weaponization.

In natural human epidemics, severe and often fatal encephalitis in Equidae (30–90 percent mortality) always precedes disease in humans. However, a biological warfare attack with virus intentionally disseminated as an aerosol would most likely cause human disease as a primary event or simultaneously with Equidae. During natural epidemics, illness or death in wild or free-ranging Equidae may not be recognized before the onset of human disease, thus a natural epidemic could be confused with a biological warfare event, and data on onset of disease should be considered with caution. A more reliable method for determining the likelihood of such an event would be the presence of VEE outside of its natural geographic range. A biological warfare attack in a region populated by Equidae and appropriate mosquito vectors could initiate an epizootic epidemic.

Medical Management

No specific viral therapy exists; hence treatment is supportive only. Patients with uncomplicated VEE infection may be treated with analgesics to relieve headache and myalgia. Patients who develop encephalitis may require anticonvulsants and intensive supportive care to maintain fluid and electrolyte balance, ensure adequate ventilation, and avoid complicating secondary bacterial infections. Patients should be treated in a screened room or in quarters treated with a residual insecticide for at least 5 days after onset, or until afebrile, as human cases may be infectious for mosquitoes for at least 72 hours. Patient isolation and quarantine is not required; sufficient contagion control is provided by the implementation of Standard Precautions augmented with the need for vector control while the patient is febrile. Patient-to-patient transmission by means of respiratory droplet infection has not been proven. The virus can be destroyed by heat (80°C for 30 minutes) and standard disinfectants.

Prophylaxis

Vaccine: There are two IND human unlicensed VEE vaccines. The first investigational vaccine (designated TC-83) was developed in the 1960s and is a live, attenuated cell-culture-propagated vaccine produced by the Salk Institute. This vaccine is not effective against all of the serotypes in the VEE complex. It has been used to protect several thousand persons against laboratory infections and is presently licensed for use in Equidae (and was used in the 1970–1971 Texas epizootic in horses), but is an IND vaccine for humans. The vaccine is given as a single 0.5 ml subcutaneous dose. Fever, malaise, and headache occur in approximately 20 percent of vaccines, and may be moderate to severe in 10 percent of those vaccines to warrant bed rest for 1–2 days. Another 18 percent of vaccines fail to develop detectable neutralizing antibodies, but it is unknown whether they are susceptible to clinical infection if challenged. Temporary contraindications for use include a concurrent viral infection or pregnancy.

A second investigational vaccine (designated C-84) has been tested but not licensed in humans and is prepared by formalin-inactivation of the TC-83 strain. This vaccine is not used for primary immunization, but is used to boost nonresponders to TC-83. Administer 0.5 ml subcutaneously at 2- to 4-week intervals for up to three inoculations or until an antibody response is measured. Periodic boosters are required. The C-84 vaccine alone does not protect rodents against experimental aerosol challenge. Therefore, C-83 is used only as a booster immunogen for the TC-84 vaccine.

As with all vaccines, the degree of protection depends upon the magnitude of the challenge dose; vaccine-induced protection could be overwhelmed by extremely high doses of the pathogen. Research is under way to produce a recombinant VEE vaccine.

Immunoprophylaxis: At present, there is no pre-exposure or post-exposure immunoprophylaxis available.

Chemoprophylaxis: In experimental animals, alpha-interferon and the interferon-inducer poly-ICLC have proven highly effective for post-exposure chemoprophylaxis of VEE. There are no clinical data on which to assess efficacy of these drugs in humans.

B.2.3 VIRAL HEMORRHAGIC FEVERS

Signs and Symptoms: Viral hemorrhagic fevers (VHFs) are feverish illnesses, which can feature flushing of the face and chest, petechiae, bleeding, edema, hypotension, and shock. Malaise, myalgias, headache, vomiting, and diarrhea may occur in any of the hemorrhagic fevers.

Diagnosis: Definitive diagnosis rests on specific virologic techniques. Significant numbers of military personnel with a hemorrhagic fever syndrome should suggest the diagnosis of a viral hemorrhagic fever.

Treatment: Intensive supportive care may be required. Antiviral therapy with ribavirin may be useful in several of these infections (available only as IND under protocol). Convalescent plasma may be effective in Argentine hemorrhagic fever (available only as IND under protocol).

Prophylaxis: The only licensed VHF vaccine is yellow fever vaccine. Prophylactic ribavirin may be effective for Lassa fever, Rift Valley fever, Congo-Crimean hemorrhagic fever, and possibly hemorrhagic fever with renal syndrome (available only as IND under protocol).

Isolation and Decontamination: Contact isolation, with the addition of a surgical mask and eye protection for those coming within three feet of the patient, is indicated for suspected or proven Lassa fever, Congo-Crimean hemorrhagic fever, or filovirus infections. Respiratory protection should be upgraded to airborne isolation, including the use of a fit-tested HEPA-filtered respirator, a battery-powered air-purifying respirator, or a positive-pressure-supplied air respirator, if patients with the above conditions have prominent cough, vomiting, diarrhea, or hemorrhage. Decontamination is accomplished with hypochlorite or phenolic disinfectants.

Overview

The viral hemorrhagic fevers are a diverse group of illnesses caused by RNA viruses from four viral families. The *Arenaviridae* include the etiologic agents of Argentine, Bolivian, and Venezuelan hemorrhagic fevers, and Lassa fever. The *Bunyaviridae* include the members of the *Hantavirus* genus, the Congo-Crimean hemorrhagic fever virus from the *Nairovirus* genus, and the Rift Valley fever virus from the *Phlebovirus* genus; the *Filoviridae* include Ebola and Marburg viruses; and the *Flaviviridae* include dengue and yellow fever viruses. These viruses are spread in a variety of ways; some may be transmitted to humans through a respiratory portal of entry. Although evidence for weaponization does not exist for many of these viruses, they are included in this manual because of their *potential* for aerosol dissemination or weaponization, or likelihood for confusion with similar agents that might be weaponized.

History and Significance

Because these viruses are so diverse and occur in different geographic locations endemically, their full history is beyond the scope of this manual. However, there are some significant events that may provide insight into their possible importance as biological threat agents.

Arenaviridae: Argentine hemorrhagic fever (AHF), caused by the Junin virus, was first described in 1955 in corn harvesters. From 300 to 600 cases per year occur in areas of the Argentine pampas. Bolivian, Brazilian, and Venezuelan hemorrhagic fevers are caused by the related Machupo, Guanarito, and Sabia viruses. Lassa virus causes disease in West Africa. These viruses are transmitted from their rodent reservoirs to humans by the inhalation of dusts contaminated with rodent excreta.

Bunyaviridae: Congo-Crimean hemorrhagic fever (CCHF) is a tick-borne disease that occurs in the Crimea and in parts of Africa, Europe, and Asia. It can also be spread by contact with infected animals, and in health care settings. Rift Valley fever (RVF) is a mosquito-borne disease that occurs in Africa. The hantaviruses are rodent-borne viruses with a wide geographic distribution. Hantaan and closely related viruses cause hemorrhagic fever with renal syndrome (HFRS) (also known as Korean hemorrhagic fever or epidemic hemorrhagic fever). This is the most common disease due to hantaviruses. It was described prior to World War II in Manchuria along the Amur River, among United Nations troops during the Korean conflict, and subsequently in Japan, China, and in the Russian Far East. Severe disease also occurs in some Balkan states, including Bosnia, Serbia, and Greece. Nephropathia epidemica is a milder disease that occurs in Scandinavia and other parts of Europe, and is caused by strains carried by bank voles. In addition, newly described hantaviruses cause Hantavirus Pulmonary Syndrome (HPS) in the Americas. The hantaviruses are transmitted to humans by the inhalation of dusts contaminated with rodent excreta.

Filoviridae: Ebola hemorrhagic fever was first recognized in the western equatorial province of the Sudan and the nearby region of Zaire in 1976. A second outbreak occurred in Sudan in 1979, and in 1995 a large outbreak (316 cases) developed in

Kikwit, Zaire, from a single index case. Subsequent epidemics have occurred in Gabon and the Ivory Coast. The African strains cause severe disease and death. It is not known why this disease appears infrequently. A related virus (Ebola Reston) was isolated from monkeys imported into the United States from the Philippines in 1989, and subsequently developed hemorrhagic fever. While subclinical infections occurred among exposed animal handlers, Ebola Reston has not been identified as a human pathogen. Marburg epidemics have occurred on six occasions: five times in Africa and once in Europe. The first recognized outbreak occurred in Marburg, Germany, and Yugoslavia, among people exposed to African green monkeys, and resulted in 31 cases and seven deaths. Filoviruses can be spread from human to human by direct contact with infected blood, secretions, organs, or semen. Ebola Reston apparently spread from monkey to monkey, and from monkeys to humans by the respiratory route. The natural reservoirs of the filoviruses are unknown.

Flaviviridae: Yellow fever and dengue are two mosquito-borne fevers that have great importance in the history of military campaigns and military medicine. Tick-borne flaviviruses include the agents of Kyanasur Forest disease in India and Omsk hemorrhagic fever in Siberia.

All of the VHF agents (except for dengue virus) are infectious by aerosol in the laboratory. These viruses could conceivably be used by an adversary as biological warfare agents, in view of their aerosol infectivity and, for some viruses, high lethality.

Medical Management

General principles of supportive care apply to hemodynamic, hematologic, pulmonary, and neurologic manifestations of VHF, regardless of the specific etiologic agent. Only intensive care

will save the most severely ill patients. Health care providers employing vigorous fluid resuscitation of hypotensive patients must be mindful of the propensity of some VHFs (for example, HFRS) for pulmonary capillary leak. Pressor agents are frequently required. The use of intravascular devices and invasive hemodynamic monitoring must be carefully considered in the context of potential benefit versus the risk of hemorrhage. Restlessness, confusion, myalgia, and hyperesthesia should be managed by conservative measures, and the judicious use of sedatives and analgesics. Secondary infections may occur as with any patient undergoing intensive care utilizing invasive procedures and devices, such as intravenous lines and indwelling catheters.

The management of clinical bleeding should follow the same principles as for any patient with a systemic coagulopathy, assisted by coagulation studies. Intramuscular injections, aspirin, and other anticoagulant drugs should be avoided.

The investigational antiviral drug ribavirin is available via compassionate use protocols for therapy of Lassa fever, HFRS, Congo-Crimean HF, and Rift Valley fever. Separate Phase III efficacy trials have indicated that parenteral ribavirin reduces morbidity in HFRS, and lowers both the morbidity and mortality of Lassa fever. In the HFRS field trial, treatment was effective if begun within the first 4 days of fever, and continued for a 7-day course. A compassionate use protocol, utilizing intravenous ribavirin as a treatment for Lassa fever, is sponsored by the CDC. Doses are slightly different, and continued for a 10-day course; treatment is most effective if begun within 7 days of onset. The only significant side effect of ribavirin is a modest anemia due to a reversible inhibition of erythropoiesis, and mild hemolysis. Although ribavirin is teratogenic in laboratory animals, the potential benefits must be weighed against the potential risks to pregnant women with grave illness due to one of these VHFs. Safety in infants and children has not been established. Ribavirin has poor *in vitro* and *in vivo* activity against the filoviruses (Ebola and Marburg) and the flaviviruses (dengue, yellow fever, Omsk HF, and Kyanasur Forest Disease).

Argentine HF responds to therapy with two or more units of convalescent plasma containing adequate amounts of neutralizing antibody and given within 8 days of onset. This therapy is investigational, and available only under protocol.

Prophylaxis

The only licensed vaccine available for any of the hemorrhagic fever viruses is yellow fever vaccine, which is mandatory for travelers to endemic areas of Africa and South America. Argentine hemorrhagic fever vaccine is a live, attenuated, investigational vaccine developed at USAMRIID, which has proved efficacious both in an animal model and in a field trial in South America, and seems to protect against Bolivian hemorrhagic fever as well. Both inactivated and live-attenuated Rift Valley fever vaccines are currently under investigation. An investigational vaccinia-vectored Hantaan vaccine is offered to laboratory workers at USAMRIID. There are currently no vaccines for the other VHF agents available for human use in the United States.

Persons with percutaneous or mucocutaneous exposure to blood, body fluids, secretions, or excretions from a patient with suspected VHF should immediately wash the affected skin surfaces with soap and water. Mucous membranes should be irrigated with copious amounts of water or saline.

Close personal contacts or medical personnel exposed to blood or secretions from VHF patients (particularly Lassa fever, CCHF, and filoviral diseases) should be monitored for symptoms, fever, and other signs during the established incubation period. A Department of Defense compassionate use protocol exists for prophylactic administration of oral ribavirin to high-risk contacts (direct exposure to body fluids) of Congo-Crimean HF patients. A similar post-exposure prophylaxis strategy has been suggested for high contacts of Lassa fever patients. Most patients will tolerate this dose well, but patients should be under surveillance for breakthrough disease (especially after drug cessation) or adverse drug effects (principally anemia).

Isolation and Containment

These viruses pose special challenges for hospital infection control. With the exception of dengue (virus present, but no secondary infection hazard) and hantaviruses (infectious virus not present in blood or excreta at the time of clinical presentation), VHF patients generally have significant quantities of virus in blood and often other secretions. Special caution must be exercised in handling sharps, needles, and other potential sources of parenteral exposure. Strict adherence to standard precautions will prevent nosocomial transmission of most VHFs.

Lassa, Congo-Crimean HF, Ebola, and Marburg viruses may be particularly prone to aerosol nosocomial spread. Secondary infections among contacts and medical personnel who were not parenterally exposed are well documented. Sometimes this occurred when the acute hemorrhagic disease (as seen in CCHF) mimicked a surgical emergency such as a bleeding gastric ulcer, with subsequent exposure and secondary spread among emergency and operating room personnel. Therefore, when one of these diseases is suspected, additional management measures are indicated. The patient should be hospitalized in a private room. An adjoining anteroom for putting on and removing protective barriers, storage of supplies, and decontamination of laboratory specimen containers should be used if available. A room with nonrecirculated air under negative pressure is advised for patients with significant cough, hemorrhage, or diarrhea. It may be wise to place the patient in such a room initially, to avoid having to transport the patient in the event of clinical deterioration. All persons entering the room should wear gloves and gowns (contact isolation). In addition, face shields or surgical masks and eye protection are indicated for those coming within three feet of the patient. Respiratory protection should be upgraded to airborne isolation, including the use of a fit-tested HEPA-filtered respirator, a battery-powered air-purifying respirator, or a positive-pressure-supplied air respirator, if patients with the above conditions have prominent cough, vomiting, diarrhea, or hemorrhage. Caution should be exercised in evaluating and treating the patient with suspected VHF. Overreaction on the part of health care providers is inappropriate and detrimental to both patient and staff, but it is prudent to provide as rigorous isolation measures as feasible.

Laboratory specimens should be double-bagged, and the exterior of the outer bag decontaminated prior to transport to the laboratory. Excreta and other contaminated materials should be autoclaved, or decontaminated by the liberal application of hypochlorite or phenolic disinfectants. Clinical laboratory personnel are also at risk for exposure and should employ a biosafety cabinet (if available) and barrier precautions when handling specimens.

No carrier state has been observed for any VHF, but excretion of virus in urine (for example, Lassa fever) or semen (for example, Argentine hemorrhagic fever) may occur during convalescence. Should the patient die, there should be minimal handling of the body, with sealing of the corpse in leak-proof material for prompt burial or cremation.

B.3 BIOLOGICAL TOXINS

Toxins are harmful substances produced by living organisms (animals, plants, microbes). Features that distinguish them from chemical agents, such as VX, cyanide, or mustard, include their not being man-made, nonvolatile (no vapor hazard), usually not dermally active (mycotoxins are the exception), and generally much more toxic per weight than chemical agents. Their lack of volatility is very important and makes them unlikely to produce either secondary or person-to-person exposures, or a persistent environmental hazard.

A toxin's utility as an aerosol weapon is determined by its toxicity, stability, and ease of production. The bacterial toxins, such as botulinum toxins, are the most toxic substances by weight known. Less toxic compounds, such as the mycotoxins, are thousands of times less toxic than botulinum, and have limited aerosol potential. Stability limits the open-air potential of some toxins. For example, botulinum and tetanus toxins are large-molecular-weight proteins, and are easily denatured by environmental factors (heat, desiccation, UV light), thus posing little downwind threat. Finally, some toxins, such as saxitoxin, might be both stable and highly toxic, but are so difficult to extract that they can only feasibly be produced in minute quantities.

As with all biological weapons, potential to cause incapacitation as well as lethality must be considered. Depending on the goals of an adversary, incapacitating agents may be more effective than lethal agents due to the overwhelming demand on the medical and evacuation infrastructure, or the expected panic in the population. Several toxins such as SEB cause significant illness at doses much lower than that required for lethality, and thus pose a significant incapacitating threat.

This section will cover four toxins considered to be among

the most likely to be used against U.S. military and civilian targets: botulinum toxins, ricin, staphylococcal enterotoxin B, and T-2 mycotoxins.

B.3.1 BOTULINUM

Signs and Symptoms: Usually begins with cranial nerve palsies, including ptosis, blurred vision, diplopia, dry mouth and throat, dysphagia, and dysphonia. This is followed by symmetrical descending flaccid paralysis, with generalized weakness and progression to respiratory failure. Symptoms begin as early as 12–36 hours after inhalation, but may take several days after exposure to low doses of toxin.

Diagnosis: Diagnosis is primarily a clinical one. Bio-warfare attack should be suspected if multiple casualties simultaneously present with progressive descending flaccid paralysis. Lab confirmation can be obtained by bioassay (mouse neutralization) of the patient's serum. Other helpful labs include: ELISA or ECL for antigen in environmental samples, PCR for bacterial DNA in environmental samples, or nerve conduction studies and electromyography.

Treatment: Early administration of trivalent licensed antitoxin or heptavalent antitoxin (IND product) may prevent or decrease progression to respiratory failure and hasten recovery. Intubation and ventilatory assistance for respiratory failure. Tracheostomy may be required.

Prophylaxis: Pentavalent toxoid vaccine (types A, B, C, D, and E) is available as an IND product for those at high risk of exposure.

Isolation and Decontamination: Standard Precautions for health care workers. Toxin is not dermally active and secondary aerosols are not a hazard from patients. Decontamination with soap and water. Botulinum toxin is inactivated by sunlight within 1–3 hours. Heat (80°C for 30 minutes, 100°C for several minutes) and chlorine (more than 99.7 percent inactivation by 3 mg/L FAC in 20 minutes) also destroy the toxin.

Overview

The botulinum toxins are a group of seven related neurotoxins produced by the spore-forming bacillus *Clostridium botulinum* and two other clostridia species. These toxins, types A through G, are the most potent neurotoxins known; paradoxically, they have been used therapeutically to treat spastic conditions (strabismus, blepharospasm, torticollis, tetanus) and cosmetically to treat wrinkles. The spores are ubiquitous; they germinate into vegetative bacteria that produce toxins during anaerobic incubation. Industrial-scale fermentation can produce large quantities of toxin for use as a biological warfare agent. There are three epidemiologic forms of naturally occurring botulism—foodborne, infantile, and wound. Botulinum could be delivered by aerosol or used to contaminate food or water supplies. When inhaled, these toxins produce a clinical picture very similar to foodborne intoxication, although the time to onset of paralytic symptoms after inhalation may actually be longer than for foodborne cases, and may vary by type and dose of toxin. The clinical syndrome produced by these toxins is known as "botulism."

History and Significance

Botulinum toxins have caused numerous cases of botulism when ingested in improperly prepared or canned foods. Many deaths have occurred secondary to such incidents. It is feasible to deliver botulinum toxins as an aerosolized biological weapon, and several countries and terrorist groups have weaponized them. Botulinum toxins were weaponized by the United States in its old offensive biological warfare program.

Evidence obtained by the United Nations in 1995 revealed that Iraq had filled and deployed over 100 munitions with nearly 10,000 liters of botulinum toxin. The Aum Shinrikyo cult in Japan weaponized and attempted to disseminate botulinum toxin on multiple occasions in Tokyo prior to their 1995 sarin attack in the Tokyo subway.

Medical Management

Supportive care, including prompt respiratory support, can be lifesaving. Respiratory failure due to paralysis of respiratory muscles is the most serious effect and, generally, the cause of death. Reported cases of botulism prior to 1950 had a mortality rate of 60 percent. With tracheotomy or endotracheal intubation and ventilatory assistance, fatalities are less than 5 percent today. Prevention of nosocomial infections is a primary concern, along with hydration, nasogastric suctioning for ileus, bowe, and bladder care, and prevention of decubitus ulcers and deep venous thromboses. Intensive and prolonged nursing care may be required for recovery, which may take up to 3 months for initial signs of improvement, and up to a year for complete resolution of symptoms.

Antitoxin: Early administration of botulinum antitoxin is critical, since the antitoxin can only neutralize the circulating toxin in patients with symptoms that continue to progress. When symptom progression ceases, no circulating toxin remains, and the antitoxin has no effect. Antitoxin may be particularly effective in foodborne cases, where presumably toxin continues to be absorbed through the gut wall. Animal experiments show that after aerosol exposure, botulinum antitoxin is very effective if given before the onset of clinical signs. If the antitoxin is delayed until after the onset of symptoms, it does not protect against respiratory failure.

Three different antitoxin preparations are available in the

United States. A licensed trivalent (types A, B, E) equine anti-toxin is available from the Centers for Disease Control and Prevention for cases of foodborne botulism. This product has all the disadvantages of a horse serum product, including the risks of anaphylaxis and serum sickness. A monovalent human antiserum (type A) is available from the California Department of Health Services for infant botulism. A "despeciated" equine heptavalent antitoxin against all seven serotypes has been prepared by cleaving the Fc fragments from horse IgG molecules. This product was developed by USAMRIID, and is currently available under IND status. It has been effective in animal studies. However, 4 percent of horse antigens remain, so there is still a risk of hypersensitivity reactions.

Use of the equine antitoxin requires skin testing for horse serum sensitivity prior to administration. Skin testing is performed by injecting 0.1 ml of a 1:10 dilution (in sterile physiological saline) of antitoxin intradermally in the patient's forearm with a 26- or 27-gauge needle. Monitor the injection site and observe the patient for allergic reaction for 20 minutes. The skin test is positive if any of these allergic reactions occur: hyperemic areola at the site of the injection >0.5 cm; fever or chills; hypotension with decrease of blood pressure >20 mm Hg for systolic and diastolic pressures; skin rash; respiratory difficulty; nausea or vomiting; generalized itching. Do NOT administer equine-derived botulinum antitoxin if the skin test is positive. If no allergic symptoms are observed, the antitoxin is administered as a single dose intravenously in a normal saline solution, 10 ml over 20 minutes.

With a positive skin test, desensitization can be attempted by administering 0.01–0.1 ml of antitoxin subcutaneously, doubling the previous dose every 20 minutes until 1.0–2.0 ml can be sustained without any marked reaction. Preferably, desensitization should be performed by an experienced allergist. Medical personnel administering the antitoxin should be prepared to treat anaphylaxis with epinephrine, intubation equipment, and IV access.

Prophylaxis

Vaccine: A pentavalent toxoid of *Clostridium botulinum* toxin types A, B, C, D, and E is available as an IND for pre-exposure prophylaxis. It will likely remain under IND status since efficacy testing in humans is not feasible. This product has been administered to several thousand volunteers and occupationally at-risk workers, and induces serum antitoxin levels that correspond to protective levels in experimental animals. The currently recommended primary series of 0, 2, and 12 weeks, followed by a 1 year booster, induces protective antibody levels in greater than 90 percent of vaccines after 1 year. Adequate antibody levels are transiently induced after three injections, but decline prior to the one-year booster.

Contraindications to the vaccine include sensitivities to alum, formaldehyde, and thimerosal, or hypersensitivity to a previous dose. Reactogenicity is mild, with 2 to 4 percent of vaccines reporting erythema, edema, or induration at the local site of injection that peaks at 24 to 48 hours. The frequency of such local reactions increases with subsequent inoculations; after the second and third doses, 7 to 10 percent will have local reactions, with higher incidence (up to 20 percent or so) after boosters. Severe local reactions are rare, consisting of more extensive edema or induration. Systemic reactions are reported in up to 3 percent, consisting of fever, malaise, headache, and myalgia. Incapacitating reactions (local or systemic) are uncommon. The vaccine should be stored at 2–8°C (not frozen).

The vaccine is recommended for selected individuals or groups judged at high risk for exposure to botulinum toxin aerosols. There is no indication at present for use of botulinum antitoxin as a prophylactic modality except under extremely specialized circumstances.

Post-exposure prophylaxis, using the heptavalent antitoxin, has been demonstrated effective in animal studies; however, human data are not available, so it is not recommended for this indication. The antitoxin should be considered for this purpose only in extraordinary circumstances.

B.3.2 RICIN

Signs and Symptoms: Acute onset of fever, chest tightness, cough, dyspnea, nausea, and arthralgias occur 4 to 8 hours after inhalational exposure. Airway necrosis and pulmonary capillary leak resulting in pulmonary edema would likely occur within 18–24 hours, followed by severe respiratory distress and death from hypoxemia in 36–72 hours.

Diagnosis: Acute lung injury in large numbers of geographically clustered patients suggests exposure to aerosolized ricin. The rapid time course to severe symptoms and death would be unusual for infectious agents. Serum and respiratory secretions should be submitted for antigen detection (ELISA). Acute and convalescent sera provide retrospective diagnosis. Nonspecific laboratory and radiographic findings include leukocytosis and bilateral interstitial infiltrates.

Treatment: Management is supportive and should include treatment for pulmonary edema. Gastric lavage and cathartics are indicated for ingestion, but charcoal is of little value for large molecules such as ricin.

Prophylaxis: There is currently no vaccine or prophylactic antitoxin available for human use, although immunization appears promising in animal models. Use of the protective mask is currently the best protection against inhalation.

Isolation and Decontamination: Standard Precautions for

health care workers. Ricin is nonvolatile, and secondary aerosols are not expected to be a danger to health care providers. Decontaminate with soap and water. Hypochlorite solutions (0.1 percent sodium hypochlorite) can inactivate ricin.

Overview

Ricin is a potent protein cytotoxin derived from the beans of the castor plant (*Ricinus communis*). Castor beans are ubiquitous worldwide, and the toxin is fairly easy to extract; therefore, ricin is potentially widely available. When inhaled as a small-particle aerosol, this toxin may produce pathologic changes within 8 hours and severe respiratory symptoms followed by acute hypoxic respiratory failure in 36–72 hours. When ingested, ricin causes severe gastrointestinal symptoms followed by vascular collapse and death. This toxin may also cause disseminated intravascular coagulation, microcirculatory failure, and multiple organ failure if given intravenously in laboratory animals.

History and Significance

Ricin's significance as a potential biological warfare toxin relates in part to its wide availability. Worldwide, one million tons of castor beans are processed annually in the production of castor oil; the waste mash from this process is 5 percent ricin by weight. The toxin is also quite stable and extremely toxic by several routes of exposure, including the respiratory route. Ricin was apparently used in the assassination of Bulgarian exile Georgi Markov in London in 1978. Markov was attacked with a specially engineered weapon disguised as an umbrella, which implanted a ricin-containing pellet into his body. This technique was used in at least six other assassination attempts in the late 1970s and early 1980s. In 1994 and 1995, four men from a tax-protest group known as the "Minnesota Patriots Council" were convicted of possessing ricin and conspiring to use it (by mixing it with the solvent DMSO) to murder law enforcement officials. In 1995, a Kansas City oncologist, Deborah Green, attempted to murder her husband by contaminating his food with ricin. In 1997, a Wisconsin resident, Thomas Leahy, was arrested and charged with possession with intent to use ricin as a weapon. **Ricin has a high terrorist potential due to its ready availability, relative ease of extraction, and notoriety in the press.**

Medical Management

Management of ricin-intoxicated patients depends on the route of exposure. Patients with pulmonary intoxication are managed by appropriate respiratory support (oxygen, intubation, ventilation) and hemodynamic monitoring) and treatment for pulmonary edema, as indicated. Gastrointestinal intoxication is best managed by vigorous gastric lavage, followed by use of cathartics such as magnesium citrate. Superactivated charcoal is of little value for large molecules such as ricin. Volume replacement of GI fluid losses is important. In percutaneous exposures, treatment would be primarily supportive.

Prophylaxis

The protective mask is effective in preventing aerosol exposure. Although a vaccine is not currently available, candidate vaccines are under development, which are immunogenic and confer protection against lethal aerosol exposures in animals. Pre-exposure prophylaxis with such a vaccine is the most promising defense against a biological warfare attack with ricin.

B.3.3 STAPHYLOCOCCAL ENTEROTOXIN B

Signs and Symptoms: Latent period of 3–12 hours after aerosol exposure is followed by sudden onset of fever, chills, headache, myalgia, and nonproductive cough. Some patients may develop shortness of breath and retrosternal chest pain. Patients tend to plateau rapidly to a fairly stable clinical state. Fever may last 2 to 5 days, and cough may persist for up to 4 weeks. Patients may also present with nausea, vomiting, and diarrhea if they swallow the toxin. Presumably, higher exposure can lead to septic shock and death.

Diagnosis: Diagnosis is clinical. Patients present with a feverish respiratory syndrome without chest X-ray abnormalities. Large numbers of patients presenting in a short period of time with typical symptoms and signs of SEB pulmonary exposure would suggest an intentional attack with this toxin.

Treatment: Treatment is limited to supportive care. Artificial ventilation might be needed for very severe cases, and attention to fluid management is important.

Prophylaxis: Use of protective mask. **There is currently no human vaccine available to prevent SEB intoxication.**

Isolation and Decontamination: Standard Precautions for health care workers. SEB is not dermally active and secondary aerosols are not a hazard from patients. Decontamination with soap and water. Destroy any food that may have been contaminated.

Overview

Staphylococcus aureus produces a number of exotoxins, one of which is staphylococcal enterotoxin B, or SEB. Such toxins are referred to as exotoxins since they are excreted from the organism, and since they normally exert their effects on the intestines they are called enterotoxins. SEB is one of the pyrogenic toxins that commonly causes food poisoning in humans after the toxin is produced in improperly handled foodstuffs and subsequently ingested. SEB has a very broad spectrum of biological activity. This toxin causes a markedly different clinical syndrome when inhaled than it characteristically produces when ingested. Significant morbidity is produced in individuals who are exposed to SEB by either portal of entry to the body.

History and Significance

SEB is the second most common source of outbreaks of food poisoning. Often these outbreaks occur in a setting such as a church picnic or other community event, due to common source exposure in which contaminated food is consumed. Although an aerosolized SEB toxin weapon would not likely produce significant mortality, it could render 80 percent or more of exposed personnel clinically ill and unable to perform their mission for 1–2 weeks. The demand on the medical and logistical systems could be overwhelming. For these reasons, SEB was one of the seven biological agents stockpiled by the United States during its old bioweapons program, which was terminated in 1969.

Medical Management

Currently, therapy is limited to supportive care. Close attention to oxygenation and hydration is important, and in severe cases with pulmonary edema, ventilation with positive end expiratory pressure, vasopressors, and diuretics might be necessary. Acetaminophen for fever and cough suppressants may make the patient more comfortable. The value of steroids is unknown. Most patients would be expected to do quite well after the initial acute phase of their illness, but generally would be unfit for duty for one to two weeks. Severe cases risk death from pulmonary edema and respiratory failure.

Prophylaxis

Although there is currently no human vaccine for immunization against SEB intoxication, several vaccine candidates are in development. Preliminary animal studies have been encouraging. A vaccine candidate is nearing transition to advanced development for safety and immunogenicity testing in humans. Experimentally, passive immunotherapy can reduce mortality in animals, but only when given within 4–8 hours after inhaling SEB. Because of the rapidity of SEB's binding with MHC receptors (<5 min *in vitro*), active immunization is considered the most practical defense. Interestingly, most people have detectable antibody titers to SEB and SEC1, however, immunity acquired through natural exposure to SEB does not provide complete protection from an aerosol challenge (although it may reduce the emetic effect).

B.3.4 T-2 MYCOTOXINS

Signs and Symptoms: Exposure causes skin pain, pruritus, redness, vesicles, necrosis, and sloughing of the epidermis. Effects on the airway include nose and throat pain, nasal discharge, itching and sneezing, cough, dyspnea, wheezing, chest pain, and hemoptysis. Toxin also produces effects after ingestion or eye contact. Severe intoxication results in prostration, weakness, ataxia, collapse, shock, and death.

Diagnosis: Should be suspected if an aerosol attack occurs in the form of "yellow rain" with droplets of variously pigmented oily fluids contaminating clothes and the environment. Confirmation requires testing of blood, tissue, and environmental samples.

Treatment: There is no specific antidote. Treatment is supportive. Soap and water washing, even 4–6 hours after exposure, can significantly reduce dermal toxicity; washing within 1 hour may prevent toxicity entirely. Superactivated charcoal should be given orally if the toxin is swallowed.

Prophylaxis: The only defense is to prevent exposure by wearing a protective mask and clothing (or topical skin protectant) during an attack. No specific immunotherapy or chemotherapy is available for use in the field.

Isolation and Decontamination: Outer clothing should be removed and exposed skin decontaminated with soap and water. Eye exposure should be treated with copious saline irrigation. Secondary aerosols are not a hazard; however, contact with contaminated skin and clothing can produce secondary dermal exposures. Contact Precautions are warranted until decontamination is accomplished. Then, Standard Precautions are recommended for health care workers. Environmental decontamination requires the use of a hypochlorite solution under alkaline conditions, such as 1 percent sodium hypochlorite and $0.1M$ NaOH with 1-hour contact time.

Overview

The trichothecene (T-2) mycotoxins are a group of over 40 compounds produced by fungi of the genus *Fusarium*, a common grain mold. They are small-molecular-weight compounds, and are extremely stable in the environment. They are the only class of toxin that is dermally active, causing blisters within a relatively short time after exposure (minutes to hours). Dermal, ocular, respiratory, and gastrointestinal exposures would be expected after an attack with mycotoxins.

History and Significance

The potential for use as a biological warfare toxin was demonstrated to the Russian military shortly after World War II when flour contaminated with species of *Fusarium* was unknowingly baked into bread that was ingested by civilians. Some developed a protracted lethal illness called alimentary toxic aleukia (ATA) characterized by initial symptoms of abdominal pain, diarrhea, vomiting, and prostration, and within days fever, chills, myalgias, and bone-marrow depression with granulocytopenia and secondary sepsis. Survival beyond this point allowed the development of painful pharyngeal/laryngeal ulceration and diffuse bleeding into the skin (petechiae and ecchymoses), melena, bloody diarrhea, hematuria, hematemesis, epistaxis, and vaginal bleeding. Pancytopenia and gastrointestinal ulceration and erosion were secondary to the ability of these toxins to profoundly arrest bone-marrow and mucosal protein synthesis and cell cycle progression through DNA replication.

Mycotoxins allegedly were released from aircraft in the "yellow rain" incidents in Laos (1975–1981), Kampuchea (1979–1981), and Afghanistan (1979–1981). It has been estimated that there were more than 6,300 deaths in Laos, 1,000 in Kampuchea, and 3,042 in Afghanistan. The alleged victims were usually unarmed civilians or guerrilla forces. These groups were not protected with masks or chemical protective clothing and had little or no capability of destroying the attacking enemy aircraft. These attacks were alleged to have occurred in remote jungle areas, which made confirmation of attacks and recovery of agent extremely difficult. Some investigators have claimed that the "yellow clouds" were, in fact, bee feces produced by swarms of migrating insects. Much controversy has centered upon the veracity of eyewitness and victim accounts, but there is some evidence to make these allegations of biological warfare agent use in these areas possible.

Medical Management

No specific antidote or therapeutic regimen is currently available. All therapy is supportive. If a person is unprotected during an attack the outer uniform should be removed within 4 hours and decontaminated by exposure to 5 percent hypochlorite for 6–10 hours. The skin should be thoroughly washed with soap and uncontaminated water if available. This can reduce dermal toxicity, even if delayed 4–6 hours after exposure. The M291 Skin Decontamination Kit can also be used to remove skin-adherent T-2. Standard burn care is indicated for cutaneous involvement. Standard therapy for poison ingestion, including the use of superactivated charcoal to absorb swallowed T-2, should be administered to victims of an unprotected aerosol attack. Respiratory support may be necessary. The eyes should be irrigated with normal saline or water to remove toxin.

Prophylaxis

Physical protection of the skin, mucous membranes, and airway (use of chemical protective mask and clothing) are the only proven effective methods of protection during an attack. Immunological (vaccines) and chemoprotective pretreatments are being studied in animal models, but are not available for field use. Topical skin protectant may limit dermal exposure. Soap and water washing, even 1 hour after dermal exposure to T-2, effectively prevents dermal toxicity.

Appendix C

CHEMICAL AGENTS AND THEIR EFFECTS

C.1 Pulmonary Agents

    C.1.1   Phosgene (CG)

C.2 Blood Agents

    C.2.1   Cyanide (AC) and Chloride (CK)

C.3 Vesicants

    C.3.1   Mustard (HD, H)

    C.3.2   Lewisite (L)

    C.3.3   Phosgene Oxime (CX)

C.4 Nerve Agents

    C.4.1   Tabun (GA)

    C.4.2   Sarin (GB)

    C.4.3   Soman (GD)

    C.4.4   GF

    C.4.5   VX

C.5 Incapacitating Agents

    C.5.1   BZ, Agent 15

C.1 PULMONARY AGENTS

C.1.1 Phosgene (CG)

Signs and Symptoms: Eye and airway irritation, dyspnea, chest tightness, and delayed pulmonary edema.

Detection: Odor of newly mown hay or freshly cut grass or corn. Neither the M256A1 detector kit nor chemical-agent detector paper (M8 paper, M9 paper) is designed to identify phosgene, but the MINICAMS, Monitox Plus, Draeger tubes, Individual Chemical Agent Detector (ICAD), M18A2, M90, and M93A1 Fox will detect small concentrations of this gas.

Decontamination: Decontamination methods vary with state: Vapor—fresh air; liquid—copious water irrigation.

Management: Termination of exposure, ABCs of resuscitation, enforced rest and observation, oxygen with or without positive airway pressure for signs of respiratory distress, other supportive therapy as needed.

Overview

Inhalation of selected organohalides, oxides of nitrogen, and other compounds can result in varying degrees of pulmonary edema, usually after a symptom-free period that varies in duration with the amount inhaled. Chemically induced, acute lung injury by these groups of agents involves a permeability defect in the blood–air barrier; however, the precise mechanisms of toxicity remain an enigma.

History and Significance

John Davy first synthesized phosgene in 1812. Subsequent development as a potential chemical warfare agent led to the first battlefield use of phosgene (in shells filled solely with phosgene) at Verdun in 1917 by Germany. Later, both sides in the conflict employed phosgene either alone or in mixed-substance shells, usually in combination with chlorine. Although military preparations for World War II included the manufacture and stockpiling of phosgene-filled munitions, phosgene was not used during the war.

Medical Management

Terminate exposure as a vital first measure. This may be accomplished by physically removing the casualty from the contaminated environment or by isolating him from surrounding contamination by supplying a properly fitting mask. Decontamination of liquid agent on clothing or skin terminates exposure from that source.

Execute the ABCs of resuscitation as required. Establishing an airway is especially critical in a patient exhibiting hoarseness or stridor; such individuals may face impending laryngeal spasm and require intubation. Steps to minimize the work of breathing must be taken.

Enforce rest. Even minimal physical exertion may shorten the clinical latency period and increase the severity of respiratory symptoms and signs in an organohalide casualty, and physical activity in a symptomatic patient may precipitate acute clinical deterioration and even death.

Prepare to manage airway secretions and prevent/treat bronchospasm. Unless super infection is present, secretions present in the airways of phosgene casualties are usually copious and watery. They may serve as an index to the degree of pulmonary edema and do not require specific therapy apart from suctioning and drainage. Antibiotics should be reserved for those patients with an infectious process documented by sputum gram staining and culture.

Prevent/treat pulmonary edema. Positive airway pressure provides some control over the clinical complications of pulmonary edema. Early use of a positive pressure mask may be beneficial. Positive airway pressure may exacerbate hypotension by decreasing thoracic venous return, necessitating intravenous fluid administration and perhaps judicious use of the pneumatic anti-shock garment.

Prevent/treat hypoxia. Oxygen therapy is definitely indicated and may require supplemental positive airway pressure administered via one of the several available devices for generating intermittent or continuous positive pressure. Intubation with or without ventilatory assistance may be required, and positive pressure may need to be applied during at least the end-expiratory phase of the ventilator cycle.

Prevent/treat hypotension. Sequestration of plasma-derived fluid in the lungs may cause hypotension that may be exacerbated by positive airway pressure. Urgent intravenous administration of either crystalloid or colloid may need to be supplemented by the judicious application of the pneumatic anti-shock garment.

C.2   BLOOD AGENTS

C.2.1 Cyanide (AC) and Chloride (CK)

Signs and Symptoms: After exposure to high enough doses, seizures, followed by respiratory and cardiac arrest.

Detection: The M250A1 detector ticket detects hydrogen cyanide (AC) as vapor or gas in the air, and the M272 kit detects cyanide in water. The ICAD, M18A2, and M90 detectors also detect cyanide. The CAM, M8A1 automatic chemical alarm, and M8 and M9 paper do not detect cyanide.

Decontamination: Skin decontamination is usually not necessary because the agents are highly volatile. Wet, contaminated clothing should be removed and the underlying skin decontaminated with water or other standard decontaminates.

Management: Intravenous sodium nitrite and sodium thiosulfate is the best antidote. Oxygen and the correct acidosis provide supportive management.

Overview

Cyanide is a rapidly acting lethal agent that is limited in its military uses by its high volatility. **Death occurs within six to eight minutes after inhalation of a high dose. Sodium nitrite and sodium thiosulfate are effective antidotes.**

History and Significance

The French used about 4,000 tons of cyanide in World War I without notable military success, possibly because the small one- to two-pound munitions used could not deliver the large amounts needed to cause biological effects. Other factors included the high volatility of cyanide (which quickly evaporated and dispersed) and its "all or nothing" biological activity, that is, it caused few effects below the lethal level (as opposed to mustard, which causes eye damage at 1 percent of the lethal amount).

Cyanides are also called "blood agents," an antiquated term still used by many in the military. At the time of the introduction of cyanide in World War I, the other chemical agents in use caused mainly local effects. Riot-control agents injured the skin and mucous membranes from direct contact, and phosgene damaged the lungs after inhalation. In contrast, inhaled cyanide produces systemic effects and was thought to be carried in the blood; hence the term "blood agent." The widespread distribution of absorbed nerve agents and vesicants via the blood invalidates this term as a specific designator for cyanide.

Medical Management

Management of cyanide poisoning begins with removal to fresh air. Dermal decontamination is unnecessary if exposure has been only to vapor, but wet clothing should be removed and the underlying skin should be washed with soap and water or water alone if liquid on the skin is a possibility. Attention to the basics of intensive supportive care is critical and includes mechanical ventilation as needed, circulatory support with crystalloids and vasopressors, correction of metabolic acidosis with IV sodium bicarbonate, and seizure control with benzodiazepine administration. The fact that cyanide inhibits cellular utilization of oxygen would lead to the expectation that supplemental oxygen would not be of use in cyanide poisoning. However, administration of 100 percent oxygen has been found empirically to exert a beneficial effect and should be a part of general supportive care for every cyanide-poisoned patient.

Symptomatic patients, especially those with severe manifestations, may further benefit from specific antidotal therapy. This is provided in a two-step process. First, a methemoglobin-forming agent such as amyl nitrite or sodium nitrite is administered. The equilibrium of this reaction causes dissociation of bound cyanide from cytochrome and frees the enzyme to help produce ATP again. The orthostatic hypotension produced by nitrite administration is not usually a concern in a severely intoxicated and prostrate cyanide casualty, but overproduction of methemoglobin may compromise oxygen-carrying capacity.

The second step in treatment is the provision of a sulfur donor, typically sodium thiosulfate, which is utilized as a substrate by rhodanese for its conversion of cyanide to thiocyanate. Sodium thiosulfate itself is efficacious, relatively benign, and also synergistic with oxygen administration and thus may be used without nitrites empirically in situations such as smoke inhalation with high carboxyhemoglobin levels.

C.3 VESICANTS

Sulfur mustard has posed a military threat since its introduction on the battlefield in World War I. Most of this section concerns this agent. Unless otherwise noted, the term "mustard" refers to sulfur mustard.

The nitrogen components (HN_1, HN_2, and HN_3) were synthesized in the 1930s, but were not produced in large amounts for warfare. Mechlorethamine (HN_2, Mustargen) became the prototypical cancer chemotherapeutic compound and remained the standard compound for this purpose for many years.

Lewisite (L) was synthesized during the late stages of World War I, but probably has not been used on a battlefield. The lewisite antidote, British-Anti-Lewisite (BAL), finds medicinal use today as a heavy-metal chelator.

Although classified as a vesicant, phosgene oxime (CX) is a corrosive urticant that also has not seen battlefield use.

Lewisite and phosgene oxime pose only minor potential military threats and will be discussed briefly at the end of this section.

C.3.1 Mustard (HD, H)

Signs and Symptoms: Asymptomatic latent period (hours). Erythema and blisters on the skin; irritation, conjunctivitis, corneal opacity, and damage to the eyes; mild upper respiratory signs to marked airway damage, also gastrointestinal (GI) effects and bone-marrow stem cell suppression.

Detection: M256A1, M272 water testing kit, MINICAMS, the ICAD, M18A2, M21 remote sensing alarm, M90, M93A1, Bubbler, CAM, and DAAMS, M8 paper or M9 paper.

Decontamination: 0.5 percent hypochlorite, M291 kit, and water in large amounts.

Management: Decontamination immediately after exposure is the only way to prevent damage. Supportive care of patients—there is no specific therapy.

Overview

Vesicant agents, specifically sulfur mustard (H, HD), have been major military threat agents since their introduction in World War I. They constitute both a vapor and a liquid threat to all exposed skin and mucous membranes. Mustard's effects are delayed, appearing hours after exposure. Organs most commonly affected are the skin (with erythema and vesicles), eyes (with mild conjunctivitis to severe eye damage), and airways (with mild irritation of the upper respiratory tract, to severe bronchiolar damage leading to necrosis and hemorrhage of the airway mucosa and musculature). Following exposure to large quantities of mustard, precursor cells of the bone marrow are damaged, leading to pancytopenia and increased susceptibility to infection. The GI tract may be damaged, and there are sometimes central nervous system (CNS) signs. **There is no specific antidote, and management is symptomatic therapy. Immediate decontamination is the only way to reduce damage.**

History and Significance

Sulfur mustard was first synthesized in the early 1800s and was first used on the battlefield during World War I by Germany in July 1917. Despite its introduction late in that conflict, mustard produced the most chemical casualties, although fewer than 5 percent of the casualties who reached medical treatment died. Italy allegedly used mustard in the 1930s against Abyssinia. Egypt apparently employed mustard in the 1960s against Yemen, and Iraq used mustard in the 1980s against Iran and the Kurds. Mustard is still considered a major threat agent of former Warsaw Pact countries, Third World countries, and terrorist groups.

Medical Management

The management of a patient exposed to mustard may be simple, as in the provision of symptomatic care for a sunburn-like erythema, or extremely complex, as in providing total management for a severely ill patient with burns, immunosuppression, and multi-system involvement. Suggested therapeutic measures for each organ system are provided below. Guidelines for general patient care are not intended to take the place of sound clinical judgment, especially in the management of complicated cases.

Skin: Erythema should be treated with calamine or other soothing lotion or cream (for example, 0.25 percent camphor and menthol, calamine) to reduce burning and itching. Small blisters should be left intact, but because larger ones will eventually break, they should be carefully unroofed. Denuded areas should be irrigated three to four times daily with saline, another sterile solution, or soapy water and then liberally covered with a topical antibiotic.

Eyes: Conjunctival irritation from a low dosage will respond to any number of available ophthalmic solutions after the eyes are thoroughly irrigated. Regular application of homatropine (or other anticholinergic drug) ophthalmic ointment will reduce or prevent future synechiae formation. A topical antibiotic applied several times a day will reduce the incidence and severity of infection.

Pulmonary: Upper airway symptoms (sore throat, nonproductive cough, and hoarseness) may respond to steam inhalation and cough suppressants. Although a productive cough and dyspnea accompanied by fever and leukocytosis occurring 12 to 24 hours after exposure may suggest a bacterial process to the clinician, he must resist the use of antibiotics for this process, which in fact is a sterile bronchitis or pneumonitis.

Gastrointestinal: Atropine (0.4–0.6 mg.), another anticholinergic drug or antiemetic, should control the early nausea and vomiting. Prolonged vomiting or voluminous diarrhea beginning days after exposure suggests direct involvement of the GI tract by severe systemic poisoning, a poor prognostic sign.

C.3.2 Lewisite (L)

Signs and Symptoms: Lewisite causes immediate pain or irritation of skin and mucous membranes. Erythema and blisters on the skin and airway damage similar to shoes seen after mustard exposures develop later.

Detection: M256A1, M272 water testing kit, MINICAMS, the ICAD, M18A2, M21 remote sensing alarm, M90, M93A1, Fox Bubbler, CAM and DAMMS, M8 paper or M9 paper.

Decontamination: M291, 0.5 percent hypochlorite, water in large amounts.

Management: Immediate decontamination; symptomatic management of lesions the same as for mustard lesions; a specific antidote (BAL) will decrease systemic effects.

Overview

Lewisite is a vesicant that damages the eyes, skin, and airways by direct contact. After absorption, it causes an increase in capillary permeability to produce hypovolemia, shock, and organ damage. Exposure to lewisite causes immediate pain or irritation, although lesions require hours to become full-blown.

Management of a lewisite casualty is similar to management of a mustard casualty, although a specific antidote, British-Anti-Lewisite (BAL), will alleviate some effects.

History and Significance

Dr. Wilford Lee Lewis first synthesized lewisite in 1918, but production was too late for its use in World War I. It has not been used in warfare, although some countries may stockpile it. Lewisite is sometimes mixed with mustard to achieve a lower freezing point of the mixture for ground dispersal and aerial spraying.

Medical Management

Early decontamination is the only way of preventing or lessening lewisite damage. Since this must be accomplished within minutes after exposure, this is self-aid rather than medical management.

The guidelines for the management of a mustard casualty will be useful. Lewisite does not cause damage to the hematopoietic organs as mustard does; however, fluid loss from the capillaries necessitates careful attention to fluid balance.

British Anti-Lewisite (BAL) was developed as an antidote for **lewisite**. There is evidence that BAL itself causes some toxicity, and the user should read the package insert carefully.

C.3.3 Phosgene Oxime (CX)

Signs and Symptoms: Immediate burning and irritation followed by wheal-like skin lesions and eye and airway damage.

Detection: M256A1, MI8A2, M90, and M93 Fox, MINI-CAMS, the ICAD, M21 remote sensing alarm, Bubbler, CAM, DAAMS, the M8A1 automatic chemical agent alarm, M8 paper or M9 paper.

Decontamination: Water in large amounts, 0.5 percent hypochlorite, M291.

Management: Immediate decontamination, symptomatic management of lesions.

Overview

Phosgene oxime (CX) is an urticant or nettle agent that causes a corrosive type of skin and tissue lesion. It is not a true vesicant since it does not cause blisters. The vapor is extremely irritating, and both the vapor and liquid cause almost immediate tissue damage upon contact. There is very scanty information available on CX.

History and Significance

There is no current assessment of the potential of CX as a military threat agent.

Medical Management

Management is supportive. The skin lesion should be managed in the same way that a necrotic ulcerated lesion from another cause would be managed.

C.4   NERVE AGENTS

C.4.1–C.4.5 Tabun (GA), Sarin (GB), Soman (GD), GF, and VX

Nerve agents are among the deadliest of chemical agents and many produce rapid symptoms. They include the G- and V-agents. Examples of G-agents are tabun (GA), sarin (GB), soman (GD), and GF. A V-agent is VX. In some countries "VF" agents are known as "A" agents.

Nerve agents can be dispersed by artillery shell, mortar shell, rocket, land mine, missile, aircraft spray, and aircraft bomb or bomblet. They could also be carried by a terrorist in the form of an aerosol tank or can.

Signs and Symptoms:

Vapor:
Small Exposure: miosis, rhinorrhea, mild difficulty breathing.

Large Exposure: sudden loss of consciousness, convulsions, apnea, flaccid paralysis, copious secretions, miosis.

Liquid on Skin:
Small to Moderate Exposure: localized sweating, nausea, vomiting, feeling of weakness.

Large Exposure: Sudden loss of consciousness, convulsions, apnea, flaccid paralysis, copious secretions.

Detection: M256A1, CAM, M8 paper, M9 paper, M8A1 and M8 alarm systems.

Decontamination: M291, M258A1, hypochlorite, large amounts of water.

Management: Administration of Mark I kits (atropine and pralidoxime chloride); diazepam in addition if the casualty is severe; ventilation and suction of airways for respiratory distress.

Overview

Nerve agents are the most toxic of known chemical agents. They are hazards in their liquid and vapor states and can cause death within minutes after exposure. Nerve agents inhibit acetylcholinesterase in tissue, and their effects are caused by the resulting excess acetylcholine.

History and Significance

Nerve agents were developed in pre–World War II Germany. Germany had stockpiles of nerve agent munitions during World War II but did not use them for reasons that are still unclear. In the closing days of the war, the United States and its allies discovered these stockpiles, developed the agents, and manufactured nerve agent munitions. The U.S. chemical agent stockpile contains the nerve agents sarin (GB) and VX.

Nerve agents are considered major military threat agents. The only known battlefield use of nerve agents was in the Iraq–Iran conflict. Intelligence analysts indicate that many countries have the technology to manufacture nerve agent munitions.

Medical Management

Management of a casualty with nerve agent intoxication consists of decontamination, ventilation, administration of the antidotes, and supportive therapy. The condition of the patient dictates the need for each of these and the order in which they are done. Decontamination is described in Chapter 6. Skin decontamination is described elsewhere in this manual. Skin decontamination is not necessary after exposure to vapor alone, but clothing should be removed because it may contain trapped vapor.

The need for ventilation will be obvious, and the means of ventilation will depend on available equipment. Airway resistance is high because of bronchoconstriction and secretions, and initial ventilation is difficult. The resistance decreases after atropine administration, after which ventilation will be easier.

Three drugs are used to treat nerve agent exposure, and another is used as pretreatment for potential nerve agent exposure. The three therapeutic drugs are atropine, pralidoxime

chloride, and diazepam. The drug pyridostigmine bromide is used in the pretreatment for nerve agents.

Atropine: Atropine is a cholinergic blocking or anticholinergic compound. It is extremely effective in blocking the effects of excess acetylcholine at peripheral muscarinic sites.

Pralidoxime chloride: Pralidoxime chloride is an oxime. Oximes attach to the nerve agent that is inhibiting the cholinesterase and break the agent–enzyme bond to restore the normal activity of the enzyme.

Diazepam: Diazepam is an anticonvulsant drug used to decrease convulsive activity and reduce the brain damage caused by prolonged seizure activity.

C.5 INCAPACITATING AGENTS

C.5.1 BZ, Agent 15

Signs and Symptoms: Mydriasis; dry mouth; dry skin; decreased level of consciousness; confusion; disorientation; disturbances in perception and interpretation; denial of illness; short attention span; impaired memory.

Detection: No field detector is available.

Decontamination: Gentle, but thorough flushing of skin and hair with water or soap and water is required. Bleach is not necessary. Remove clothing.

Management: Antidote: physostigmine. Supportive: monitoring of vital signs, especially core temperature.

Overview

BZ is a glycolate anticholinergic compound related to atropine; scopolamine, and hyoscyamine. Dispersal would be as an aerosolized solid (primarily for inhalation) or as agent dissolved in one or more solvents for ingestion or percutaneous absorption. Acting as a competitive inhibitor of acetylcholine at postsynaptic and postjunctional muscarinic receptor sites in smooth muscle, exocrine glands, autonomic ganglia, and the brain, BZ decreases the effective concentration of acetylcholine seen by receptors at these sites. **Thus, BZ causes peripheral nervous system (PNS) effects that in general are the opposite of those seen in nerve agent poisoning. Central nervous system effects include stupor, confusion, and confabulation with concrete and panoramic illusions and hallucinations.**

History and Significance

The use of chemicals to induce altered states of mind dates to antiquity and includes the use of plants such as thornapple, which contains combinations of anticholinergic alkaloids. The use of nonlethal chemicals to render an enemy incapable of fighting dates back to at least 600 B.C., when the Greek leader Solon had his soldiers throw hellebore roots into streams supplying water to enemy troops, who then developed diarrhea. In 184 B.C., Hannibal's army used belladonna plants to induce disorientation, and the Bishop of Muenster in A.D. 1672 attempted to use belladonna-containing grenades in an assault on the city or Groningen. In 1881, members of a railway surveying expedition crossing Tuareg territory in North Africa ate dried dates that tribesmen had deliberately contaminated with *Hyoscyamus falezlez*. In 1908, 200 French soldiers in Hanoi became delirious and experienced hallucinations after being poisoned with a related plant.

Medical Management

The admonition to protect oneself first may be difficult when dealing with any intoxication involving a latent period, since initially asymptomatic exposure to health care providers may already have occurred during the same time frame in which patients were exposed. Protection of medical staff from already absorbed and systemically distributed BZ in a patient is not needed.

General supportive management of the patient includes decontamination of skin and clothing, confiscation of weapons and related items from the patient, and observation. Physical restraint may be required in moderately to severely affected patients. The greatest risks to the patient's life are (1) injuries from his or her own erratic behavior and (2) hyperthermia, especially in patients who are in hot or humid environments or are dehydrated from overexertion or insufficient water intake. A severely exposed patient might be comatose with serious cardiac arrhythmias and electrolyte disturbances. Management of heat stress assumes a high priority in these patients. Because of the prolonged time course in BZ poisoning, consideration should always be given to evacuation to a higher echelon of care.

Appendix D

NATIONS POSSESSING WEAPONS OF MASS DESTRUCTION

The Department of Defense estimates that as many as 26 nations may possess chemical agents and/or weapons and an additional 12 may be seeking to develop them. The Central Intelligence Agency reports that at least 10 countries are believed to possess or be conducting research on biological agents for weaponization.

Significantly, the nations listed in this Appendix are those that combine a robust program of developing weapons of mass destruction (WMD) with a government that would appear to be willing to use them or sell them to terrorist groups.

Iraq

As this book goes to press, the United States and the United Nations are trying to resolve Baghdad's intransigence regarding weapons of mass destruction in accordance with UN

Resolution 1441. What follows is some history prior to the resumption of UN inspections per this resolution.

Baghdad had refused since December 1998 to allow UN inspectors into Iraq as required by Security Council Resolution 687 and subsequent Council resolutions, and inspections began again in late 2002 only at U.S. insistence. Furthermore, Iraq has engaged in extensive concealment efforts and has probably used the period since it refused inspections to attempt to reconstitute prohibited programs. Without UN-mandated inspectors in Iraq, assessing the current state of Iraq's WMD and missile programs is difficult.

Saddam's repeated publicized exhortations to his "Nuclear Mujahidin" to "defeat the enemy" added to the CIA's concerns that since the Gulf War Iraq has continued research and development work associated with its nuclear program. A sufficient source of fissile material remains Iraq's most significant obstacle to being able to produce a nuclear weapon. The intelligence community is concerned that Baghdad is attempting to acquire materials that could aid in reconstituting its nuclear weapons program.

Iraq continues to develop short-range ballistic missile (SRBM) systems that are not prohibited by the United Nations and is expanding to longer-range systems. Pursuit of UN-permitted ballistic missiles allows Baghdad to improve technology and infrastructure that could be applied to a longer-range missile program. The appearance of four Al Samoud SRBM transporter-erector-launchers (TELs) with airframes at the 31 December 2000, Al Aqsa parade indicates that this liquid-propellant missile program is nearing deployment. Two new solid-propellant "mixing" buildings at the al-Mamoun plant—the site originally intended to produce Badr-2000 (that is, Condor) solid-propellant missiles—appear especially suited to house large, UN-prohibited mixers of the type acquired for the Badr-

Table D.1

Countries Possessing or Suspected of Possessing Nuclear Weapons

Known to Possess	Suspected or Seeking
United States of America	Iraq
Russia	Iran
Ukraine	North Korea
France	Libya
United Kingdom	Algeria
People's Republic of China	South Africa
India	Israel
Pakistan	

2000 program. In fact, the CIA finds no logical explanation for the size and configuration of these mixing buildings other than an Iraqi intention to develop longer-range, prohibited missiles (that is, to mix solid propellant exclusively geared for such missiles). In addition, Iraq has begun reconstructing the "cast and cure" building at al-Mamoun, which contains large and deep casting pits that were specifically designed to produce now-proscribed missile motors.

If economic sanctions against Iraq were lifted, Baghdad probably would increase its attempts to acquire missile-related items from foreign sources, regardless of any future UN monitoring and continuing restrictions on long-range ballistic missile programs. With substantial foreign assistance and an accommodating political environment, Baghdad could flight-

Table D.2

Countries Possessing or Suspected of Possessing Chemical Weapons

Known to Possess	Suspected or Seeking
United States of America	People's Republic of China
Russia	North Korea
France	Eygpt
Libya	Israel
Iraq	Ethiopia
Iran	Taiwan
Syria	Burma

test an MRBM by mid-decade. In addition, Iraq probably retains a small, covert force of Scud ballistic missiles, launchers, and conventional, chemical, and biological warheads. The CIA assesses that, since December 1998, Iraq has increased its capability to pursue chemical warfare (CW) programs. After both the Gulf War and Operation Desert Fox in December 1998, Iraq rebuilt key portions of its chemical production infrastructure for industrial and commercial use, as well as former dual-use CW production facilities and missile production facilities. Iraq has attempted to purchase numerous dual-use items for, or under the guise of, legitimate civilian use. Since the suspension of UN inspections in December 1998, the risk of diversion of such equipment has increased. In addition, Iraq appears to be installing or repairing dual-use equipment at CW-related facilities. Some of these facilities could be converted fairly quickly for production of CW agents.

The UN Special Commission (UNSCOM) reported to the Security Council in December 1998 that Iraq also continued to withhold information related to its CW program. For example, Baghdad seized from UNSCOM inspectors an Iraqi Air Force document discovered by UNSCOM that indicated that Iraq had not consumed as many CW munitions during the Iran–Iraq war in the 1980s as had been declared by Baghdad. This discrepancy indicates that Iraq may have hidden an additional 6,000 CW munitions.

Baghdad continues to pursue a BW program. Iraq in 1995 admitted to having an offensive BW program, but UNSCOM was unable to verify the full scope and nature of Iraq's efforts. UNSCOM assessed that Iraq was maintaining a knowledge base and industrial infrastructure that could be used to produce quickly a large amount of BW agents at any time. In addition, Iraq has continued dual-use research that could improve BW agent R&D capabilities. In light of Iraq's growing industrial self-sufficiency and the likely availability of mobile or covert facilities, the CIA is concerned that Iraq may again be producing BW agents.

Iraq is pursuing an unmanned aerial vehicle (UAV) program that converts L-29 jet trainer aircraft originally acquired from Eastern Europe. In the past, Iraq has conducted flights of the L-29, possibly to test system improvements or to train new pilots. The CIA suspects that these refurbished trainer aircraft have been modified for delivery of chemical or, more likely, biological warfare agents.

North Korea

North Korea has continued attempts to procure technology worldwide that could have applications in its nuclear program.

The restarting of its Yongbyon facility and its potential for producing weapons-grade plutonium is troubling to say the least. The North has been seeking centrifuge-related materials in large quantities to support a uranium enrichment program. It also obtained equipment suitable for use in uranium feed and withdrawal systems.

North Korea claims it is in possession of several nuclear weapons. As of this printing, these claims are unconfirmed. We do know that the spent fuel rods canned in accordance with the 1994 Agreed Framework contain enough plutonium for several nuclear weapons.

North Korea also has continued procurement of raw materials and components for its ballistic missile programs from various foreign sources, especially through North Korean firms based in China. North Korea continues to abide by its voluntary moratorium on flight tests, which it has said it would observe until at least 2003.

Iran

Iran is vigorously pursuing programs to produce indigenous weapons of mass destruction—nuclear, chemical, and biological—and their delivery systems. To this end, it seeks foreign materials, training, equipment, and know-how that have enabled it to produce some complete weapon systems, with their means of delivery, and components of other weapons. Iran has focused particularly on entities in Russia, China, North Korea, and Europe.

Despite Iran's status in the Treaty on the Nonproliferation of Nuclear Weapons (NPT), the United States is convinced that Tehran is pursuing a nuclear weapons program. To bolster its efforts to establish domestic nuclear fuel-cycle capabilities, Iran has sought assorted foreign fissile materials and technology.

Such capabilities also can support fissile material production for Tehran's overall nuclear weapons program.

Despite the Bushehr power plant being put under International Atomic Energy Agency (IAEA) safeguards, Russia's provision of expertise and manufacturing assistance has enabled Iran to develop its nuclear technology infrastructure—which, in turn, can benefit directly Tehran's nuclear weapons R&D program. In addition, Russian entities continued associations with Iranian research centers on other nuclear fuel-cycle activities.

Iran has attempted to use its civilian nuclear energy program, which is quite modest in scope, to justify its efforts to establish domestically or otherwise acquire assorted nuclear fuel-cycle capabilities. Such capabilities, however, are well suited to support fissile material production for a weapons program, and the CIA believes it is this objective that drives Iran's efforts to acquire relevant facilities. For example, Iran has sought to obtain turnkey facilities, such as a uranium conversion facility (UCF) that ostensibly would be used to support fuel production for the Bushehr power plant. But the UCF could be used in any number of ways to support fissile material production needed for a nuclear weapon—specifically, production of uranium hexafluoride for use as a feedstock for uranium enrichment operations and production of uranium compounds suitable for use as fuel in a plutonium production reactor. In addition, the CIA suspects that Tehran is interested in acquiring foreign fissile material and technology for weapons development as part of its overall nuclear weapons program.

Facing economic pressures, some Russian entities have shown a willingness to provide assistance to Iran's nuclear projects by circumventing their country's export laws. Enforcement of export control laws has been inconsistent and ineffective, but the U.S. government continues to engage the Russian government in a cooperative export control dialogue. For example, an institute subordinate to the Russian Ministry of Atomic Energy (MINATOM) had agreed to deliver in late 2000 equipment that was clearly intended for atomic vapor laser isotope separation, a technology capable of producing weapons-grade uranium. As a result of U.S. protests, the Russian government halted the delivery of some of this equipment to Iran.

China is completing assistance on two Iranian nuclear projects: a small research reactor and a zirconium production facility at Esfahan that will enable Iran to produce cladding for reactor fuel. As a party to the NPT, Iran is required to accept IAEA safeguards on its nuclear material. The IAEA's Additional Protocol requires states to declare production of zirconium fuel cladding and gives the IAEA the right of access to resolve questions or inconsistencies related to the declarations, but Iran has made no moves to bring the Additional Protocol into force. Zirconium production, other than production of fuel cladding, is not subject to declaration or inspection.

Ballistic missile-related cooperation from entities in the former Soviet Union, North Korea, and China over the years has helped Iran move toward its goal of becoming self-sufficient in the production of ballistic missiles. Such assistance during the reporting period has included equipment, technology, and expertise. Iran, already producing Scud short-range ballistic missiles (SRBMs), is in the late stages of developing the Shahab-3 medium-range ballistic missile (MRBM). In addition, Iran publicly has acknowledged the development of follow-on versions of the Shahab-3. It originally said that another version, the Shahab-4, is a more capable ballistic missile than its predecessor but later characterized it as solely a space launch vehicle with no military applications. Iran's Defense Minister has also publicly mentioned a "Shahab-5." Such statements strongly suggest that Tehran intends to develop a longer-range ballistic missile capability.

Iran is a party to the Chemical Weapons Convention (CWC). Nevertheless, it continued to seek chemicals, production technology, training, and expertise from entities in Russia and China that could further efforts at achieving an indigenous capability to produce nerve agents. Iran already has stockpiled blister, blood, and choking agents—and the bombs and artillery shells to deliver them—which it previously has manufactured. It probably also has made some nerve agents.

Foreign dual-use biotechnical materials, equipment, and expertise, primarily, but not exclusively, from entities in Russia and Eastern Europe, continued to feature prominently in Iran's procurement efforts. Such materials have legitimate uses, but Iran's biological warfare (BW) program also could benefit from them.

Libya

An NPT party with full-scope IAEA safeguards, Libya continues to develop its nuclear infrastructure. The suspension of UN sanctions has provided Libya the means to enhance its nuclear infrastructure through foreign cooperation and procurement efforts. Tripoli and Moscow continued talks on cooperation at the Tajura Nuclear Research Center and a potential power reactor deal. Such civil-sector work could present Libya with opportunities to pursue technologies that also would be suitable for military purposes. In addition, Libya participated in various technical exchanges through which it could try to obtain dual-use equipment and technology that could enhance its overall technical capabilities in the nuclear area. In 2001, Libya and other countries reportedly used their secret services to try to obtain technical information on the development of weapons of mass destruction, including nuclear weapons. Although Libya is making political overtures to the West in an attempt to strengthen relations, Libya's continuing interest in nuclear weapons and ongoing nuclear infrastructure upgrades raise concerns.

The suspension of UN sanctions in 1999 has allowed Libya to expand its efforts to obtain ballistic missile–related equipment, materials, technology, and expertise from foreign sources. Outside assistance—particularly from Serbian, Indian, Iranian, North Korean, and Chinese entities—has been critical to its ballistic missile development programs. Libya's capability probably remains limited to its Scud B missiles, but with continued foreign assistance it will probably achieve an MRBM capability—a long-desired goal—or extended-range Scud capability.

Libya remains heavily dependent on foreign suppliers for CW precursor chemicals and other key related equipment. Following the suspension of UN sanctions, Tripoli reestablished contacts, primarily in Western Europe, with sources of expertise, parts, and precursor chemicals abroad. Tripoli still appears to be working toward an offensive CW capability and eventually indigenous production. Evidence suggests that Libya also is seeking to acquire the capability to develop and produce BW agents.

Syria

Syria—an NPT signatory with full-scope IAEA safeguards—has a nuclear research center at Dayr Al Hajar. Russia and Syria have approved a draft cooperative program on cooperation on civil nuclear power. In principal, broader access to Russian expertise provides opportunities for Syria to expand its indigenous capabilities, should it decide to pursue nuclear weapons. During the second half of 2001, Damascus continued to receive help from abroad on establishing a solid-propellant rocket motor development and production capability. Syria's liquid-propellant missile program has and will continue to depend on essential foreign equipment and assistance—primarily from North Korean entities and Russian firms. Damascus also continued its efforts to assemble—probably with considerable North Korean assistance—liquid-fueled Scud C missiles.

Syria sought CW-related precursors and expertise from foreign sources during the reporting period. Damascus already holds a stockpile of the nerve agent sarin but apparently is trying to develop more toxic and persistent nerve agents. Syria remains dependent on foreign sources for key elements of its CW program, including precursor chemicals and key production equipment. It is highly probable that Syria also is developing an offensive BW capability.

Sudan

Sudan, a party to the CWC, has been developing the capability to produce chemical weapons for many years. It historically has obtained help from foreign entities, principally in Iraq. Sudan may be interested in a BW program as well.

India

The underground nuclear tests in May 1998 were a significant milestone in India's continuing nuclear weapons development program. Since the 1998 tests, New Delhi has continued efforts intended to lead to the development of more sophisticated nuclear weapons. During this reporting period, India continued to obtain foreign assistance for its civilian nuclear power program, primarily from Russia.

India still lacks engineering or production expertise in some key missile technologies. Entities in Russia and Western Europe remained the primary conduits of missile-related and dual-use technology transfers during 2001. India flight-tested the Dhanush ballistic missile, continued work with the Russians on the Brahmos cruise missile, and moved nuclear-capable Prithvi missiles and launchers together within range of Pakistan as part of its military mobilization.

Pakistan

Pakistan's nuclear weapons tests in late May 1998 demonstrated its well-developed nuclear weapons program. Pakistan has continued to acquire nuclear-related equipment, some of it dual use, and materials from various sources—principally in Western Europe. If Pakistan chooses to develop more advanced nuclear weapons, seeking such goods will remain important. China provided extensive support in the past to Islamabad's nuclear weapons and ballistic missile programs, but in May 1996 it pledged not to provide assistance to unsafeguarded nuclear facilities in any state, including Pakistan. The possibility of continued contacts between Chinese and Pakistani entities on Pakistani nuclear weapons development cannot be ruled out.

Pakistan's ballistic missile program continued to benefit from significant Chinese entity assistance during the reporting period. With Chinese entity assistance, Pakistan is moving toward serial production of solid-propellant SRBMs, such as the Shaheen-I and Haider-I. Pakistan last conducted ballistic missile flight tests in 1999, but recently flight-tested the Haider-I ballistic missile in 2002. Successful development of the two-stage Shaheen-II MRBM will require continued assistance from Chinese entities or other potential sources.

Appendix E

NUCLEAR, BIOLOGICAL, AND CHEMICAL WARFARE REFERENCES

Most books of this type provide the reader with a long list of references for further reading about the subject. I originally intended to include a bibliography and began compiling a list. In short order this list became impossibly long. Additionally, the most up-to-date references in the area of nuclear, biological, and chemical warfare and terrorism are on the Internet. Simply entering any major search engine with the right combination of words will take the reader to a wealth of source material on this subject. The key—as with any search—is to narrow and focus the search in such a way that you locate relevant and reliable information.

That said, there are a number of Web sites maintained by government agencies and other entities that contain a wealth of information on the subject of nuclear, biological, and chemical warfare and terrorism and will give the reader a head-start in this area:

U.S. Army Medical Research Institute for Infectious Diseases (USAMRIID)
www.usamriid.army.mil

Military's Medical Online Information Server
www.nbc-med.org/ie40/default.html

Centers for Disease Control
www.bt.cdc.gov

World Health Organization
www.who.int/home-page

Army Field Manuals on Weapons of Mass Destruction
www.globalsecurity.org/wmd/library/policy/army/fm

Tactical Gear Command
www.tacticalgearcommand.com

Information on Gas Masks
www.bobandjulie.com/gasmask.html

Red Cross Information Web Sites
www.redcross.org
www.redcross.org/pubs/#aid
www.redcross.org/services/disaster/beprepared/supplies.html
www.redcross.org/services/disaster/beprepared/forchildren.html

Los Angles Community Emergency Response Team (CERT) Information
www.cert-la.com/manuals/instman.htm

The key to using all of these—and other Web sites that you will find on your own—is to extract the relevant information *before* a disaster strikes. A terrorist attack using weapons of mass destruction will likely overload the Internet and it will be too late to pull down this information when you are trying to react to an attack with nuclear, radiological, biological, or chemical weapons. Collect that good information *now* and write it down in a place where you and your loved ones can easily get to it.

Index